The Road to Transport Management

2nd edition

P Fawcett, FCIT, MBIM

A textbook for candidates for the Certificate of Professional Competence (National Goods and National PSV Operators)

a
FLEETBOOKS
publication

To my wife

FRANCES

and to my daughters

ANDREA and JANE

P.F.

'The Road to Transport Management' by P Fawcett
2nd edition revised and enlarged 1984

ISBN 0 86047 203 5

This book has been prepared with a view to giving the correct position on legal and other matters as at 1st August 1984. However, whilst every effort has been made to ensure accuracy, neither the author nor the publishers accept any responsibility or liability for any loss or damage resulting from any error contained in the text of this book.

Published by A M Witton (Fleetbooks)
Room 20 City Buildings
69 Corporation Street
MANCHESTER M4 2DE.

Foreword

The efficiency of the road transport industry depends upon its continued ability to make a healthy living through satisfying the wants of its customers. Ever since commercial motor transport established itself, it has been criticised for lacking the professionalism of the railway industry which it has largely supplanted, which is why those who have the interest of the industry at heart have welcomed the legislation that requires its managers to demonstrate their professional competence. The Certificate that readers of this book seek to acquire is not just a passport to a career in management, but evidence of status in the community, and thus a matter of pride. It should be remembered that it is also the first rung on the ladder of professional education, for many who have obtained their Certificate of Professional Competence have gone on to take courses for membership of one of the professional bodies in the transport industry, which may in turn offer admission to further and higher education.

In the coming years the transport industry is going to be in great need of well educated managers who also know what it is like to work at the 'sharp end'. Nowhere is this more true than in road transport, both freight and passenger. I welcome Mr Fawcett's primer, already well known and successful as it is, to a Second Edition, and wish every success in their endeavour to all those who will use it.

Dr John Hibbs, FCIT, MInstTA,
Director of Transport Studies,
City of Birmingham Polytechnic. April 1984

Introduction

The idea of writing this book came to me in 1977 as I sat on the working party of the RSA Examinations Board to prepare a draft syllabus for the Certificate of Professional Competence. As soon as the book was published, it became evident that it fulfilled the needs of National CPC (Goods Vehicle Operators) examination candidates. Over 8000 copies of the book have been sold, some in conjunction with the updating supplement which it was necessary to write in 1981. I am often asked to recommend textbooks by National CPC (Passenger Vehicle Operators) students, and by students of such diverse topics as Industrial Relations, Transport Law, Road Freight and Road Passenger Transport Operation, and Transport Finance, Costing and Accounts, studying for RSA, CIT, NEBSS and Institute of Transport Administration examinations. Accordingly I have added to this second edition an additional chapter to cover the Passenger Vehicle Operators' syllabus. If this small volume meets these students' needs at a price which they can afford, I will be satisfied.

I have tried to make the text interesting and readable, which has meant relegating many finer points to footnotes, but it is my intention that the serious student should find in the footnotes all the references necessary to pursue each subject further. For ease of reference, footnotes are given at the foot of each page rather than, as hitherto, at the end of chapters. One footnote which I have not included is an injunction to the reader to treat as synonymous all references to 'The Minister', 'The Minister of Transport Industries' or 'The Secretary of State', since the Acts and Regulations now in force originated variously with the former Ministry of Transport, the Department of the Environment, or the Department of Transport. (Where the last-named body is mentioned in the text it is usually abbreviated to 'DTp'). References in Chapter 12 to the Secretary of State relate of course to the Secretary of State at the Department of Employment.

I am grateful to the RSA Examinations Board for permission to reproduce the syllabus for the CPC as Appendix 3 of this book. Appendix 4 consists of 24 questions of my own invention, intended to give the reader some idea of the form and standard of the CPC Examination. Appendices 1 and 2 contain a Table of Statutes and a Table of Cases respectively; these features have been introduced for the first time in the present edition.

The law as stated is, as far as I can ascertain, correct as at 1st August 1984. However, if any reader detects any factual errors, either I or my publisher will be grateful for a note of these. This will enable us to make any necessary corrections in subsequent editions.

Finally, I can do no better than to wish all candidates attempting the CPC exam, success in their endeavours; and to remind them, as first edition readers were reminded in the Foreword to that book by Mr A F Beckenham, then Director of Education and Training at the Chartered Institute of Transport, that "the passing of the statutory test could be the beginning for many to pursue higher studies in Transport Management."

P FAWCETT, FCIT, MBIM,
Senior Lecturer in Transport,
Faculty of Business Studies,
St John's Centre,
Central Manchester College,
Lower Hardman Street,
MANCHESTER, M3 3FP.

Manchester
July 1984

Contents

Diagrams

1 Operators' Licensing

The reader of this book will probably want to qualify to hold, or to be named on, a Goods Vehicle or PSV Operator's Licence. Throughout chapters 1-12 the assumption is made that the reader is a Goods Vehicle Operator; much of this material also applies to PSV Operators, but the latter should turn to chapter 13 (starting on page 165) for those matters where PSV practice varies from that for Goods Operators. Chapters 1-2 explain the theory and practice of licensing for Goods Operators. The book itself aims to impart the facts which need to be known in order to qualify.

1.1 General Principles of Licensing

The Government issues various types of licences to permit (1) the holder to engage in specific activities. These activities the Government can thus monitor and control, for some or all of the following reasons:-

(a) to ensure an acceptable level of safe operation;
(b) to protect the environment;
(c) to prevent wasteful competition (2);
(d) to maintain essential public services (3);
(e) to protect licensed operators from unfair competition (2)(3);
(f) to raise revenue through licence fees (4).

There is theoretically more than one way of controlling road freight transport. For example, it is possible to license the operator, the vehicles, the traffic (5) or the routes (6). A Government's pricing policy may also be seen as an indirect form of licensing (7).

---

(1) The Public Passenger Vehicles Act 1981 actually refers to a modified form of PSV Minibus Operator's Licence as a 'Permit'.
(2) This was one of the ostensible reasons for the system of Goods Vehicle Licensing in the Road Traffic Acts 1933/1960, viz.:-
'A' Licence - Hire and Reward
'C' Licence - Own Account Operation
'B' Licence - Limited Hire and Reward plus Own Account.
The Geddes Commission (1967) recommended the disbandment of this system of <u>quantity licensing</u>, and it was replaced by the <u>quality licensing</u> provisions of the Transport Act 1968.
(3) One of the stated reasons for PSV Road Service Licences (Road Traffic Act 1930).
(4) e.g. Vehicle Excise Licences.
(5) As described above in footnote (2).
(6) e.g. PSV Road Service Licences.
(7) e.g. the use of fuel taxes and subsidies as economic regulators.

1.1.1 Quality and Quantity Licensing

Of the above reasons for licensing, (a) and (b) imply a control over the quality of operations, while (c) to (f) inclusive imply a control over the quantity of operations.

The only method of applying a system of quality licensing to road freight transport operations is to license the operator to operate specified vehicles. Operators' Licensing is such a system of quality licensing. It is based purely on considerations of safe operation, and takes no account of the economic or commercial aspects (i.e. the quantity aspects).

1.2 Operators' Licensing

An Operator's Licence (referred to as an O-licence) is required by any person using a goods vehicle on the road for hire or reward, or in connection with a trade or business (8), with certain exceptions noted below.

A person using a goods vehicle in contravention of this provision is liable to a fine not exceeding £500.

1.2.1 Scope of O-Licensing

With certain exceptions, the user of a goods vehicle is required to hold an O-licence. Exemptions (9) from O-licensing mainly relate to vehicles which only incidentally carry goods, such as agricultural machinery, some specialised vehicles owned by local authorities, and dual-purpose vehicles. However, the most important exemption is for 'small goods vehicles'.

A 'small goods vehicle' is defined in the Transport Act 1968 as a vehicle which:-
(a) is under 3.5 tonnes Gross Plated Weight (10) (or, if unplated, is under 30 cwts. (1525kg) Unladen Weight); or
(b) if it is a motor vehicle and trailer combination, both of which are plated, the sum of the Gross Plated Weights of both is less than 3.5 tonnes (10)(11); or
(c) if it is a motor vehicle and trailer combination, either of which is unplated, the sum of their Unladen Weights is less than 30 cwts. (1525kg); or
(d) if it is an articulated combination with a plated semi-trailer, the sum of the Gross Plated Weight of the semi-trailer and the Unladen Weight of the tractive unit is less than 3.5 tonnes (10); or
(e) if it is an articulated combination with an unplated semi-trailer, the sum of the Unladen Weights of the combination's parts is less than 30 cwts. (1525kg).

The user of a goods vehicle can be defined (12) as the driver (if the vehicle belongs to him (13)), or the person whose servant or agent the driver is. This means that an owner-driver who works for an operator must himself hold the vehicle's O-licence; but if an operator arranges for his employee to drive the vehicle, then the operator must hold the O-licence (14).

---

(8) Transport Act 1968, section 60.
(9) Goods Vehicles (O-Licences, Qualifications and Fees) Regulations 1984, schedule 5.
(10) For vehicles first used before 1st January 1977, read 3.5 tons.
(11) The weight of any trailer under 1 ton unladen is discounted.
(12) Transport Act 1968, section 92; Goods Vehicles (Operators' Licences, Qualifications and Fees) Regulations 1984, section 3.
(13) Or is in his possession under a loan or hire agreement.
(14) Ready Mixed Concrete (East Midlands) Ltd v. Yorkshire Traffic Area Licensing Authority, Transport Tribunal 1970.

1.2.2 Applications for O-licences

Applications must be made to the Licensing Authority (LA), who is the Chairman of the Traffic Commissioners for the Traffic Area in which the vehicles are to be based. Thus operators with operating centres in more than one Traffic Areas will need to hold more than one O-licence.

An operator may apply for a licence to cover vehicles in his possession, and vehicles which he intends to acquire during the currency of the licence. (The latter is often referred to as his 'margin').

The application must specify vehicles, trailers and semi-trailers used by the applicant (15). Discs are issued for vehicles specified (but not for trailers, semi-trailers, or vehicles authorised but not specified).

Variations If an additional vehicle is required to be used, or if a vehicle is substituted (16) for a licensed vehicle, application must be made for a variation of the licence. Additions can also be allowed for within the 'margin', in which case the LA need only be notified within one month of the addition.

Publication Details of new licence applications and significant variations are published by the LA (17), as well as details of any Public Inquiries.

Press Notice In addition, an applicant must publish in a local newspaper details of his application in a prescribed form (18). The notice must be published during a period of 21 days either side of the date of the application. If this is not done, the LA must refuse the application.

The Transport Acts 1968 and 1982 (19) list the information which the LA may require an applicant to give, and this is fully considered under 1.2.4 below. The LA may also ask for any further information he may require.

1.2.3 Objections and Appeals

Objections may be made in writing to the LA in the prescribed form (20), not more than three weeks after publication of an application or proposed variation. Objectors have to provide supporting evidence and send a copy of their objection to the applicant. No objection can be made on commercial grounds, but there are a number of grounds given in the Transport and Road Traffic Acts (21), any of which if upheld would cast doubt on the quality or safety of operation of the vehicle(s).

The 1968 Act (22) lays down a list of statutory objectors, reproduced below. The right of an individual to object, other than through one of these organisations, is effectively curtailed.

---

(15) A semi-trailer or trailer serial number allocated by the Driver and Vehicle Licensing Centre, Swansea, will suffice.
(16) The old licence disc must be returned and a new disc obtained.
(17) Applications and Decisions, fortnightly, available on subscription from local Traffic Area Offices of the DTp.
(18) Transport Act 1982, section 52 and schedule 4.
(19) Transport Act 1968, section 62, as added to and amended by Road Traffic Act 1974, schedule 4; Transport Act 1982, section 52 and schedule 4.
(20) Goods Vehicles (O-Licences, Qualifications and Fees) Regulations 1984, section 11.
(21) Transport Act 1968, section 64; Road Traffic Act 1974, schedule 4; Transport Act 1982, section 52 and schedule 4. See also section 1.2.4 in this chapter.
(22) Transport Act 1968, section 63(3), as amended by Goods Vehicles (O-Licences, Qualifications and Fees) Regulations 1984, section 17.

The LA must refuse the application if he is satisfied as to the validity of the objection, but the burden of proof is on the objector. If the LA decides to grant the application in spite of an objection, he may either grant it in full or in a modified form, e.g. he may forbid any addition to the licence, vary the type or number of vehicles applied for, or remove an operating centre.

The statutory grounds for objection are considered fully in 1.2.4 below.

Statutory Objectors
- The Police
- The Local Authority
- Planning Authorities
- British Association of Removers
- Freight Transport Association
- Road Haulage Association
- General and Municipal Workers' Union
- Transport and General Workers' Union
- National Union of Railwaymen
- United Road Transport Union
- Union of Shop, Distributive and Allied Workers

Currency of the Licence Licences are normally issued for a period of five years, but they may be issued for longer or shorter periods to suit the LA's administrative convenience. The LA may also issue a licence for less than five years if he thinks the situation appropriate (e.g. to allow an operator to remain in business while the LA's examiners make an investigation and report back to him). If necessary, the LA may prematurely terminate a licence (23).

Appeals An applicant aggrieved by the failure of the LA to grant his application for either (a) an O-licence or (b) a Certificate of Professional Competence (24), or an objector to (a) above who is aggrieved by the grant of a licence, may appeal to the Transport Tribunal.

The LA also has disciplinary powers of suspension, revocation or curtailment of (a) above (25). He may hold a Public Inquiry at which the holder will be required to show reason why the licence should not be so dealt with. An LA may also give notice to the holder of a Standard Licence that he is considering its revocation, if it appears to him that the holder no longer satisfies the requirements as to Good Repute, Financial Standing or Professional Competence, allowing the holder 21 days to make a written representation (26). The holder may similarly appeal to the Transport Tribunal against the LA's action in regard to his licence.

The Acts (25) note the matters which the LA may consider when deciding whether to suspend, revoke or curtail a licence (23), and these are considered in full in 1.2.4 below.

1.2.4 Matters which the LA is required to consider

These will be dealt with under broad subject headings, but it should be noted from the foregoing that the Acts (25) give four lists of inter-related considerations, viz:-

---

(23) The LA has no power to suspend part of a licence under section 92. His proper course is to curtail the licence for a shorter period than the remainder of its duration. (Transport Tribunal 1971, re Kelman of Turriff's appeal).
(24) This aspect is dealt with in full in Chapter 2.
(25) Transport Act 1968, section 69; Road Traffic Act 1974, schedule 4; Transport Act 1982, section 52, schedule 4.
(26) The LA must consider any duly made representation (Goods Vehicles (O-Licences, Qualifications and Fees) Regulations 1984, section 9).

Transport Act 1968, section 62: Information which the LA may require;
Transport Act 1968, section 64: Statutory grounds for objection;
Transport Act 1968, section 69: Considerations relevant to the suspension, revocation or curtailment of a licence;
Transport Act 1982, section 52: Environmental considerations.

Maintenance The LA may require details about the number and type of vehicles and trailers which it is proposed to operate, and the purpose for which they will be used. He will also want to know about the arrangements and facilities for maintaining the vehicles in a fit and serviceable condition (27). While the DTp lays down no rigid guidelines about maintenance facilities, it does point out that their adequacy or otherwise will be reflected by the condition of the operator's vehicles on the road (28). Even if the maintenance arrangements are contracted out to a garage, the operator can only delegate the authority to carry out maintenance on his behalf; he is still responsible for ensuring that the vehicles are properly maintained. Where maintenance is thus sub-contracted, it is usual to provide for the LA's benefit a 'statement of intent' from the contractor to the operator.

It should also be noted that the LA need only consider whether, if the licence is granted, there will be satisfactory facilities and arrangements - not whether there might be such at some future date (29).

Financial Resources The LA may require details about the financial resources available to the applicant. 'Being of sufficient financial standing' consists of having available sufficient financial resources to ensure the establishment and proper administration of the undertaking to be licensed (30). At a Public Inquiry the LA may ask the applicant to supply a banker's reference, and the LA may be assisted by an assessor appointed by the Minister (31).

The Inquiry will be held in camera if the applicant so wishes.

A statutory ground of objection is that the facilities and arrangements for maintenance will be prejudiced by insufficient financial resources. The applicant must show that he finances his maintenance out of working capital, and not out of revenue (32)(33).

The LA must also take into consideration the fact that the holder of the licence has been adjudicated bankrupt or gone into liquidation (34) when deciding whether to suspend, revoke or curtail the licence.

'Fit Person' It is sufficient grounds for objection to show that the applicant is not a 'fit person' to hold an O-licence. This is similar to the requirement of the EEC Directive on Access to the Occupation of Road Goods Operator (35) that the operator should be of 'good repute'; and indeed, the O-licence provisions of

(27) Transport Act 1968, section 62.
(28) DTp 'Guide for Goods Vehicle Operators', Ref. GV74.
(29) Transport Tribunal re B H King & Son (Transport) Ltd. In this case the operator had planning permission for new premises.
(30) Goods Vehicles (O-Licences, Qualifications and Fees) Regulations 1984, schedule 6.
(31) Transport Act 1968, section 64.
(32) Transport Tribunal re Bullen's appeal, 1971.
(33) An operator who employs only employee drivers to drive his vehicle is not thereby 'ipso facto' an unfit person to hold a licence, nor does it follow that he will necessarily have inadequate financial resources to maintain them. Transport Tribunal 1984, re Kammac Trucking and North Western LA.
(34) Transport Act 1968, section 69.
(35) EEC 561/74.

the 1968 Act enable the UK to comply with the Directive. Obviously, a history of previous convictions will go to establish that the applicant is not a fit person; but the Act goes further than this, in attempting to stop an operator who is 'unfit' from continuing in business by proxy or in another guise. The LA may require the applicant, for this purpose, to disclose details of past and present activities both of himself (if an owner-driver) and/or the company and its directors, involving the use of the vehicles. In the case of a company, the LA may require details of the directors and officers of the company, and of any holding company, and of any person, with whom the vehicles are to be operated in partnership. Thus if there is any material change in the licence holder's circumstances, or if the person whose licence is suspended or revoked is a director or partner in, or holds a controlling position in, another firm, or occupies a similar position in the parent company of such a firm, that firm's licence may also be suspended or revoked (33).

Operating arrangements The LA may require details of the arrangements which the operator makes to ensure:-

(a) that he complies with the legislation relating to drivers' hours and rest (36);

(b) that he does not operate his vehicles so that they are overloaded (37);

The LA may, for example, depending on the nature of the traffic, be interested to know if the operator has a weighbridge or reasonable access to one, or if his vehicles have self weighing devices.

It is a statutory ground of objection to show that these arrangements will not be satisfactory.

The LA may also consider whether the operating centre is suitable, with sufficient off-street parking (38). The operating centre is redefined by the Transport Act 1982 to mean the base or centre at which the vehicle is normally kept (rather than, as under previous legislation, normally used) (39).

Environmental Considerations The bodies listed above who can object to the grant or variation of a licence are now able (40) to object on the additional grounds that an operating centre is unsuitable on environmental grounds, insofar as the operations there will be capable of prejudicially affecting the use or enjoyment of the land. Persons who own or occupy land in the **vicinity** of an operating centre are also able to make representations to the LA against grant or variation of an O-licence. Objections or representations have to be made in writing to the LA, with a copy being sent to the applicant. Timing is important; objections must be lodged within 21 days after the appearance of the application in Applications and Decisions, representations within 21 days after the appearance of the press notice. Also, whilst an objector has the right of appeal to the Transport Tribunal (see Appeals, above), a representor does not. The onus of proving the matter alleged rests with the objector or representor.

The LA may refuse a licence on environmental grounds, or he may attach conditions to the licence which he thinks appropriate to prevent or minimise environmental

---

(36) Transport Act 1968, Part VI.

(37) Road Traffic Act 1974 section 160; Construction and Use Regulations 1978, Schedule 7, as amended by Construction and Use (Amendment) (No.7) Regulations, 1982.

(38) Road Traffic Act 1974, schedule 4.

(39) Transport Act 1982, Section 52, reversing the decision in A Cash & A McCall v. North West Traffic Area Licensing Authority, Transport Tribunal 1976.

(40) Transport Act 1968, section 62, as added to and amended by Road Traffic Act 1974, schedule 4; Transport Act 1982, section 52 and schedule 4.

intrusion. He may also, where more than one operating centre is applied for, specify only those centres which he regards as suitable, and either refuse the others or attach conditions to them. However, if an existing operator can satisfy the LA that the renewal of his licence would not result in any **material change**, the LA cannot refuse the licence (except on the grounds that parking of the vehicles in the vicinity would adversely affect the environment), and can only impose conditions if he first gives the applicant the opportunity to explain the possible effect of these conditions on his business. He must then give "special consideration" to such an explanation.

In deciding the suitability of an operating centre on environmental grounds, the LA must be guided by the following **considerations** (40). Note that he has no independent right of investigation, and may only consider these matters if he receives an objection or representation:-

(a) the nature and use of any other land in the vicinity of the operating centre;
(b) the extent to which a licence grant would result in a material change to a previous or existing operating centre, or its use, which would adversely affect the environment in the vicinity;
(c) any planning application or permission relating to a new operating centre or other land in its vicinity;
(d) the number, type and size of authorised vehicles;
(e) arrangements for parking authorised vehicles;
(f) the nature and times of use of the land as an operating centre;
(g) the nature and times of use of any equipment at the operating centre; and
(h) the means and frequency of vehicles' ingress to, and egress from, the operating centre.

The conditions which the LA can impose when the licence is granted or varied materially are given below. They can be attached whether or not objections or representations have been received, and varied or removed at any time (41):-

(a) the number, type and size of authorised vehicles which may be at an operating centre at any one time for maintenance (including fuelling) or parking;
(b) the parking arrangements to be provided for authorised vehicles at or in the vicinity of each operating centre;
(c) the times between which maintenance or movement of an authorised vehicle may be carried out at an operating centre, or the times at which any equipment may be used for maintenance or moving vehicles; and
(d) the means of ingress to, and egress from, an operating centre for any authorised vehicles.

'Section 69 Offences' In most editions of Applications and Decisions there will be found, under the heading 'Public Enquiries for consideration under Section 69', a list of defaulters whom the LA has called before him to 'show reason why their licences should not be suspended or revoked'.

Applicants are also required to furnish the LA with particulars of any previous convictions by themselves or their designated Managers (42) during the last **five** years for any of the offences specified under Section 69.

The specified offences are those which the LA may take into consideration when considering whether to suspend, revoke or curtail a licence. The list has grown since the Transport Act 1968, and the LA may now also consider any offence which arises between the date on which the application is made, and the date on which

---

(40) Transport Act 1968, section 62, as added to and amended by Road Traffic Act 1974, schedule 4; Transport Act 1982, section 52 and schedule 4.
(41) Transport Act 1968, new section 69C, inserted by the Goods Vehicles (O-Licences, Qualifications and Fees) Regulations 1984.
(42) Under Section 65 relating to 'Transport Managers' Licences'.

the LA disposes of it (43).

The matters which the LA may consider under Section 69 as amended and extended (other than those already dealt with above) are:-

(a) that the holder made a false statement of fact or intention in order to obtain an O-licence;
(b) that there has been a relevant material change in the holder's circumstances (e.g. bankruptcy);
(c) that during the preceding five years there has been a conviction arising from contravention of any of the following:-
  (i) regulations relating to speed limits, vehicle weights (loaded and unloaded) and drivers' licensing;
  (ii) regulations relating to the operation of licensed goods vehicles (44);
  (iii) Part VI of the Act relating to drivers' hours and records;
  (iv) Customs and Excise Act 1952 relating to the unlawful use of rebated fuel;
  (v) regulations requiring the maintenance of vehicles in a fit and serviceable condition (45) and the keeping of records of inspection of goods vehicles (46);
  (vi) regulations relating to prohibitions of use of unfit goods vehicles (47);
  (vii) Plating and Testing regulations (48);

  or that there have been frequent convictions relating to:-
  (i) Lorry Route Restrictions (48) (49);
  (ii) Waiting, parking, loading and unloading regulations.

Finally, the LA may require details of the designated full-time Transport Manager named under Section 65; it is grounds for objection that the proposed person is unsatisfactory, and the LA may take into consideration under Section 69 any contravention of Section 65. The above provision will not apply to restricted O-licences. This is the subject of Chapter 2.

---

(43) Road Traffic Act 1974, Schedule 4.
(44) Transport Act 1968, Part V.
(45) Construction and Use Regulations 1978.
(46) Road Traffic Act 1972, Section 59.
(47) Road Traffic Act 1972, Section 57 (i.e. GV9 Procedure).
(48) Addition to list made by Road Traffic Act 1974, Schedule 4.
(49) Heavy Commercial Vehicles (Control and Regulation) Act 1973.

2 Access to the Occupation

There is an obvious shortcoming in the O-licence system as described in Chapter 1. This is that, where the Licensing Authority is faced with an application from an operator about whom he knows nothing (often for the very good reason that it is a new application), the LA has great difficulty in judging the applicant's suitability. He can call for bank references and, if relevant, copies of maintenance agreements, but without any past record on which to judge the applicant, he must grant the licence while instructing his staff to monitor the operations, by visiting the premises, inspecting vehicles, records and tachograph discs, and reporting any defects or irregularities for consideration under Section 69.

2.1 Background to the Legislation

Legislation to control admission to the occupation of Road Haulier was seen as a necessity for some time after O-licensing was introduced, because of the relative ease with which unsuitable persons could enter the industry.

2.1.1. The 'Transport Manager's Licence'

The Act of 1968 (1) made provision for there to be named on every O-licence a Licensed Transport Manager, and set out in some detail how this was to be achieved (2). However, the 'TML', as it became known, was never implemented.

The provisions did in fact relate to all operators, whether or not they carried for hire and reward, and their purpose was purely to strengthen the quality (safety) licensing provisions of the Act, by providing for revocation of a TML for infringement of the O-licence criteria.

2.1.2 The EEC Directive

The Directive (3) placed an obligation on member governments to implement such a system by 1st January 1978. It differed from the 1968 Act (1) in two important ways:-

(a) The Directive, while it covers carriage by road of freight and passengers (4), relates only to hire and reward activities.

(b) The Directive goes further than the Act insofar as it is concerned not only with the operator's ability to work safely within the law, but also with his commercial acumen - in particular his ability to remain commercially sound, and to understand those aspects of social and company law which affect his operations.

---

(1) Transport Act 1968, section 65.
(2) Transport Act 1968, schedule 9.
(3) EEC Directive 561/74.
(4) The 1968 Act made no provision for licensing Passenger Transport Managers.

2.2 Access to the Occupation of Road Haulier

2.2.1 The Goods Vehicles (O-Licences, Qualifications and Fees) Regulations 1984

Regulations made in 1977 (5) extended the O-licence provisions of the Transport Act 1968 so as to make the UK comply with the EEC Directive from 1st January 1978.

These are now consolidated in the above-mentioned 1984 Regulations, whose effect is as follows.

2.2.2 Scope

For the purpose of implementing the Directive, Operators' Licences were divided into two classes from 1st January 1978.

(a) Standard O-licence - applicable to carriers operating for hire and reward, i.e. Road Hauliers and some Own Account Operators. A Standard O-licence is endorsed to indicate whether it may be used
(i) for domestic operation only; or
(ii) for domestic and international operation.

(b) Restricted O-licence - applicable to Own Account Operators for the carriage of goods in connection with the holder's trade or business (6).

2.2.3 Sanctions

It is an offence (maximum fine £500) to use a Restricted O-licence for hire and reward, and if there is a second offence within 5 years, the LA must revoke the licence.

It is an offence (maximum fine £500) to use a domestic Standard O-licence for international hire and reward operation, and the LA has power to revoke the licence in such cases.

2.2.4 Professional Competence

Holders of Restricted O-licences will only need to satisfy the conditions set out in the Transport Act 1968 (7). However, the applicant for a Standard O-licence will in addition have to satisfy the LA respecting:-
(a) his/her financial standing;
(b) his/her good repute;
(c) his/her professional competence.

The requirement as to 'good repute' is similar to the Transport Act's (8) requirement for the LA to satisfy himself as to the applicant being a 'fit person' to hold an O-licence, by reference to his record (if any) of relevant convictions during the last five years. However, by coupling this with a requirement to have appropriate financial standing, the new Regulations go further. In effect this

---

(5) The Goods Vehicle Operators (Qualifications) Regulation 1977.
(6) An early DTp consultative paper proposed making Section 65 of the Transport Act 1968 the instrument for implementing the EEC Directive. This would have effectively brought into the scope of the regulations all own-account operators. However, this expedient has never been employed by the DTp, because of the administrative cost.
(7) As extended by the Road Traffic Act 1974, schedule 4, and the Transport Act 1982, section 52, schedule 4.
(8) Transport Act 1968, section 64.

means not simply that the applicant has not become bankrupt or gone into liquidation (9), or that his maintenance might be prejudiced by lack of finance (8), but that he has sufficient finances to safely carry on his operations.

The professional competence requirement can be met by an individual licence-holder, or by a firm which employs one or more professionally competent Transport Managers (10). ONLY INDIVIDUALS CAN BE PROFESSIONALLY COMPETENT. If the applicant has more than one operating centre and if the LA so requires, the requirement may have to be met by the appointment of more than one <u>full-time</u> qualified Transport Manager (11). All Transport Managers specified on a Standard International O-licence should hold an International CPC, even if their depot carries out no International work (12).

A condition that the holder of the CPC named on the O-licence must be resident in the same traffic area as the licence's Operating Centre is unnecessary if the professionally competent person can effectively control operations from his base (13).

2.2.5 Derogations

Where the requirements of the Regulations can be met in some alternative way - either indefinitely, or for a limited time so that an operator can continue in business while he takes the necessary steps to comply fully - the Regulations take account of such situations by granting 'derogations' as follows.

(a) 'Grandfather Rights' Individuals who had held an O-licence, or were in responsible road transport employment, before 1st January 1975, were regarded as professionally competent. 'Responsible employment' meant that the person had held an O-licence, either individually or with others, or had been responsible for the operation of Goods Vehicles used under an O-licence.

Individuals had until 31st December 1979 to apply on this basis to the Licensing Authority for a certificate. This was then valid for a Standard O-licence for both domestic and international operations, even if the applicant's pre-1975 operations were either

(i) purely domestic; or
(ii) exclusive of hire and reward work.

The Licensing Authority had powers to grant or refuse any application for a Certificate of Professional Competence on this basis. However, once granted, even

---

(8) Transport Act 1968, section 64.
(9) Transport Act 1968, section 69: 'Matters which the LA may consider in deciding whether to suspend, curtail or revoke a licence'.
(10) The same professionally competent Transport Manager cannot be specified on licences held by two separate companies (North Western Licensing Authority re Go Fast Shipping 1979; West Midlands Licensing Authority re G Seedhouse Transport Ltd 1978). The Transport Manager must be FULL TIME and not "a retired professionally competent friend" (North Western Licensing Authority re G J Pilling 1978) or an "independent contractor" (Transport Tribunal, Scotflow v. Eastern Licensing Authority, 1969).
(11) The Transport Act 1968 provided for one Transport Manager for each Operating Centre, or conversely to allow more than one Operating Centre to be controlled by one Transport Manager where this would be practical, as in North Western Licensing Authority v. Tootle's Textiles 1978.
(12) Pickford's Removals v. Metropolitan Deputy Licensing Authority, 1981 ('Motor Transport', 5th August 1981).
(13) Scot Bowyers Ltd v. South Wales Licensing Authority, Transport Tribunal 1979.

if it was not immediately used, the professionally competent person may retain it until required.

(b) New entrants from 1st January 1978

In all other cases the individual requires a Certificate of Professional Competence issued by a body approved for the purpose by the Secretary of State.

The Royal Society of Arts has been confirmed as the examining body for the Certificate of Professional Competence, and they provide examinations in the September, December, March and June of each academic year.

Revocation of a Standard O-licence - in consequence of the death, or physical or legal incapacity of:-
(i) the holder of an O-licence; or
(ii) his designated Transport Manager required by the licence.

The licence may continue in force for up to 18 months at the LA's discretion (14).

This is dealt with by section 10 of the Regulations (15), the essential parts of which are set out in the Table on page 26.

2.2.6 Miscellaneous Provisions

(a) Variation The holder of a Restricted O-licence may apply to convert this into a Standard O-licence, and the holder of a Domestic Standard O-licence may apply to convert this to an International Standard O-licence, if they can comply with the additional requirements as to Professional Competence (in the latter case by obtaining an International CPC) (16).

(b) Subsidiaries Where the holder of a Standard O-licence is a Holding Company and the vehicles specified thereon belong to its Subsidiary Company, they will be treated as if belonging to the Holding Company (17).

(c) Weight Threshold The Regulations apply to the operation of Goods Vehicles over 3.5 tonnes Gross Plated Weight (18).

(d) Form of the examination Part I consists of a one and a half hour's paper containing 60 objective test (19) questions relating to the operation of Domestic Road Haulage services. Part II is an optional one-hour paper on International Haulage, containing 30 objective test (19) questions.

(e) Syllabus for the Domestic Examination
Safety (including Drivers' Hours and Records, Driving Licences, and Road Traffic Acts and Regulations) - 15 hours' directed study.
Technical Standards (including Mechanical Condition, Construction and Use, Weights and Dimensions, Plating and Testing) - 11 1/2 hours' directed study.

---

(14) Goods Vehicles (O-Licences, Qualifications and Fees) Regulations 1984, schedule 6(5).
(15) Ibid., section 10.
(16) Ibid., section 8.
(17) Ibid., section 32.
(18) Transport Act 1968, section 60.
(19) The questions will call for the minimum of writing by the candidate, whose answers will be indisputably right or wrong.

Access to the Market (O-Licensing and Professional Competence) - 8 hours' directed study.

Business and Financial Management (including Costing, Insurance, Management, Law of Business, Carriage and Employment) - 13 1/2 hours' directed study.

(f) Other Qualifications An individual will be considered professionally competent if he or she can satisfy the LA that he/she holds an **alternative qualification** recognised for the purpose by the Secretary of State. A comprehensive list of these is published by the Department of Transport in their 'Guide to Goods Vehicle Operators' Licences'.

(g) Appeals An applicant aggrieved by the failure of the LA to grant his application for a Certificate of Professional Competence, or by the LA's revocation of his Certificate, may appeal to the Transport Tribunal.

2.3 Professionalism and Standards of Operation

Even owner-drivers, if they wish to carry for hire or reward on a Standard O-licence, will need to be professionally competent. Many operators have in the past damaged the industry's reputation, albeit in the short term before the LA has been able to apply any sanctions, by their use of unsafe vehicles operating at uneconomic and unfair rates. They will now either be 'weeded out' by the regulations or, as a result of the training which they must undertake in pursuit of the Certificate, they will commence sound, commercial and safe operations. Demand for professionally competent persons has occurred within the industry, bringing to the Transport Manager the enhanced status and remuneration which the occupation rightly deserves.

EVENTS AFFECTING THE VALIDITY OF A CERTIFICATE OF PROFESSIONAL COMPETENCE
(Figures in brackets indicate subsections of the Goods Vehicles (O-licences, Qualifications and Fees) Regulations 1984, section 10 'Derogations')

Nature of Event	Effect on Standard O-Licence	Effect on Restricted O-Licence
5(a) Licence held or application made by an individual who (i) dies, (ii) becomes bankrupt or (iii) becomes a mental patient; 5(b) Licence held or application made by a corporate body which (i)(ii) is compulsorily or voluntarily wound up, (iii) is put into receivership, or (iv) has its property possessed by debenture holders; 5(c) Licence held or application made by a partnership which is (i) dissolved or (ii) a partner becomes a mental patient, leaving a sole partner in the firm	THE LICENCE CEASES TO HAVE EFFECT, or the application fails, except as below:- (6)(a) Notice is given to the LA within **2 months** (i) that the licence holder ceases to be the user and (ii) the name of the person to whom the business has been transferred and (iii) the successor applies for a Standard O-licence within **4 months** or a Restricted O-Licence within **1 month**	Effect as for Standard O-licence
5(a)(iv) 5(b)(v) 5(c)(iii) The requirement as to professional competence ceases to be satisfied	The LA need not revoke the licence during a period of up to **1 year**, or, if he so determines, during a further period of up to **6 months**	No effect on Restricted O-Licences
An O-licence extended as above will continue in force provided that:-	(7) It will not continue beyond the original expiry date, and (8) The 'caretaker holder' is subject to the same provisions as the original applicant regarding convictions, activities of applicant and directors (Transport Act 1968 section 62(4)(d)(e))	

3 Excise Duty and Driving Licences

3.1 Vehicle Excise Duty and Registration

The operator of a Goods Vehicle or PSV becomes liable for the payment of a number of fees to Government departments in respect of his Operator's Licence, Drivers' Licences, Plating and Testing, Fuel Tax and Excise Duty.

3.1.1 Vehicle Excise Acts 1962 and 1971

Taxation of vehicles, in its narrowest sense, is the Vehicle Excise Licensing Duty, commonly referred to as Road Fund Tax, which is payable in respect of all mechanically propelled vehicles used on a public road, i.e. a road maintained at the public expense (Vehicles (Excise) Act 1971) (1). Note that this definition of a 'public road' is much narrower than the definition given in the Road Traffic Acts, which is given primarily for Road Traffic Offences purposes.

Vehicle Excise Licence Duty is payable to the Vehicle Licensing Office (2) which acts as an agent of the Exchequer in collecting fees and issuing licences. (The Road Fund was merged with the Exchequer in 1936). All mechanically propelled vehicles, before being used for the first time on a public road, require to be registered with the Vehicle Licensing Office. In order to do this the appropriate form V55, which is obtainable at Money Order Post Offices or from the Local Vehicle Licensing Office (LVLO), must be submitted, either to the LVLO or to the Goods Vehicle Centre at Swansea, together with

(a) the appropriate Licence Fee;
(b) a valid certificate of insurance for the vehicle (3);
(c) evidence of the Unladen Weight in the case of a goods vehicle (i.e. a weighbridge ticket);
(d) evidence that the vehicle is a new vehicle, or that since its manufacture it has not been used on a public road;
(e) a Certificate of Conformity or Minister's Approval Certificate.

In addition form V205 is required where a rigid vehicle exceeding 12,000 kg Gross Vehicle Weight is to be used with a draw-bar trailer exceeding 4,000 kg Gross Weight.

The Vehicle Licensing Office will then issue the vehicle with a Registration

---

(1) Section 8. Note that under section 5, no offence is committed if an untaxed vehicle is driven to or from a pre-booked annual roadworthiness test.
(2) Vehicle and Driving Licences Act 1969; Vehicle and Driving Licences (Transfer of Functions) Order 1971. The Local Vehicle Licensing Offices (LVLOs) perform these functions locally as agents of the Department of Transport's Driver and Vehicle Licensing Centre (DVLC) at Swansea.
(3) Motor Vehicles (Third Party Risks) Regulation 1961, and Road Traffic Act 1972, section 153.

Document (Form V5), commonly called a 'Log Book'. Note that the holder of the 'Log Book' may not necessarily be the owner of the vehicle. For example, under a hire purchase agreement the hirer will hold the 'Log Book', having possession of the vehicle, but not owning it. Also, under the Hire Purchase Act 1965, if such a hirer sold the vehicle while it was in his possession to a private person, that person could acquire a good title to the vehicle, i.e. become the owner. (The HP company could then sue the hirer for conversion!)

At registration a licence may be obtained for 12 or 6 months, and may subsequently be renewed for similar periods. In the case of a 6 month licence, the fee payable is 11/20ths of the annual rate. If it is desired to renew the licence at a Head Post Office this can be done provided that application is made either
(a) not more than 14 days before the licence becomes effective; or
(b) not more than 14 days after the previous licence has expired
and that there has been no change in the vehicle itself, its use, or its ownership. In all other cases the renewal must be made at the Local Vehicle Licensing Office or through the DVLC (2). The application form V11, whether presented at a Head Post Office or at the LVLO, or posted, must be accompanied by:
(a) and (b) licence fee and certificate of insurance as above;
(c) (i) where the vehicle is subject to Plating and Testing, the relevant test certificate (4); or
(ii) where the vehicle is subject to annual test after three years under the 1960 Road Traffic Act, the relevant test certificate (5).

The licence disc issued for the vehicle must be displayed (6) on the nearside lower corner of the windscreen, so that all particulars are clearly visible from the road (7).

If the vehicle is not going to be used for a complete month or multiples thereof, and the licence disc is surrendered, a <u>**refund**</u> amounting to 1/12th of the annual fee for each month can be claimed.

If the vehicle is disposed of, the fact must be reported to the Local Vehicle Licensing Office giving the name and address of the new owner, who must then transfer the registration.

Vehicles which use public roads only for the purposes of passing between premises or land owned by the owner of the vehicle may be exempt from Vehicle Excise Duty by special application to the Local Vehicle Licensing Office, provided that the distance which the vehicle travels on the public road is less than 6 miles per week.

It is an offence to use or keep an unlicensed mechanically propelled vehicle on a public road (8). Even if the engine is removed from a motor vehicle, it does not cease to be a mechanically propelled vehicle (9).

The Vehicle and Driving Licences Act 1969 is stricter still in making it an offence to keep a mechanically propelled vehicle on a road for no matter how short a time without duty being paid. Therefore, technically it is an offence to keep an unlicensed vehicle on a public road even during the 'Days of Grace', which, although not a legal provision, were 14 days allowed at the beginning of

---

(4) Goods Vehicles (Production of Test Certificates) Regulation 1970.
(5) Motor Vehicles (Test) (Extension) Order 1966.
(6) Transport Act 1968, section 147.
(7) To fail to do so is an offence. Vehicles (Excise) Acts 1962 and 1971.
(8) Vehicles going to and from annual roadworthiness tests can be exempted from Vehicle Excise Duty only if a definite appointment has been made.
(9) Newberry v. Simmonds 1961.

each month for renewing a licence disc - provided that the old disc, which had expired in the immediately previous month, is still displayed.

Although the DVLC may take proceedings against a driver for using or keeping an unlicensed vehicle on a public road, the courts would prefer the action to be taken against the owner of the vehicle, where he is the employer of the driver (10). The competent authority to bring a prosecution is the DVLC, but proceedings may be instituted by a Police Constable duly authorised to do so (11).

3.1.2 Rates of Duty - Finance Acts

The rate of duty payable on a vehicle depends upon the manner in which the vehicle is constructed, and the manner in which it is used. The rate payable may be altered by the Government, and the Finance Act appropriate to the most recent change in Vehicle Excise Duty should be consulted to establish the actual taxation rate. However, the Vehicle Excise Acts give much useful guidance on the classification of vehicles for taxation purposes.

A mechanically propelled vehicle will be taxed at the **Private Rate** if it is a private car or a passenger vehicle (other than a Public Service Vehicle which is taxed as a 'Hackney Carriage' on the basis of seating capacity). Even if the car or passenger vehicle is used to carry business goods, it may still be taxed at the private rate provided it has not been adapted to carry goods (12).

A vehicle constructed or adapted to carry goods may nevertheless be taxed at the private rate if an undertaking is given to the DVLC that it will not be used to carry goods. A training vehicle carrying only ballast to simulate loaded conditions may also be taxed at the private rate.

The Goods Vehicle Rate is payable if the vehicle is constructed or adapted to carry goods (i.e. if it is a Goods Vehicle by definition) AND if it is used to carry goods for hire or reward, or in connection with a trade or business. The rate is based upon the Gross Plated Weight and number of axles of the vehicle. However, Goods Vehicles of up to 30 cwts Unladen Weight now form a separate taxation class, Private Light Goods Vehicles, for which a flat rate is payable.

In calculating the Unladen Weight of the vehicle for tax purposes the weight of the fuel carried, water, any loose tools or covers and ropes, and batteries if the vehicle is battery-electric (the batteries being counted as fuel) can be ignored. There is no obligation to carry a spare wheel, but if one is carried it must be included in the Unladen Weight. Generally, containers and demountable bodies are excluded from the Unladen Weight of a vehicle; but demountable bodies or containers so constructed that they constitute an alternative body, will be included in the Unladen Weight (13), which will have to be calculated on the basis of the heavier of the alternative bodies.

Trailers Trailer supplement is payable on trailers exceeding 4,000 kg Gross Weight, towed by goods vehicles with a Gross Plated Weight exceeding 12,000 kg, taxed as goods vehicles, where the trailer itself is used to carry goods for hire or reward, or in connection with a trade or business. Trailer supplement is an additional duty which is based on the Gross Plated Weight of the trailer - not the drawing vehicle.

---

(10) Carpenter v. Campbell 1953.
(11) Vehicles (Excise) Act 1971. The wording of the Act would suggest that such an authority must be specific to a particular case, and not merely general.
(12) Flower Freight Co v. Hammond, 1962.
(13) Paterson v. Burnet 1939. The test is whether they can be lifted on or off when loaded.

__Articulated Vehicles__ are treated as a single vehicle for taxation purposes. This means that where a number of semi-trailers are towed by a mechanically propelled vehicle, the rate of tax will have to be computed on the basis of the Train Weight of the towing vehicle and, where this exceeds 12 tonnes Gross Train Weight, either:-

(a) a basic scale for units used with a mixture of 1-axle, 2-axle or 3-axle semi-trailers, or

(b) one of two concessionary scales where the operator undertakes not to use a semi-trailer with more than one or two axles respectively.

__'Down Licensing'__ The Finance Act 1982 (14), which introduced Vehicle Excise Duty based on Gross Plated Weight, contained a provision enabling the Secretary of State to institute an 'Operating Weight' for taxation purposes (which would be less than the Gross Plated Weight, and would be noted on the licence and not necessarily on the Plate). An operator would undertake not to exceed this weight in operation. There are obvious advantages for the carriers of light bulky loads who may require a vehicle with a high design weight, to assist in keeping schedules. The taxation implications are obvious.

Contravention of the 'operating weight' would be an offence resulting in the higher rate of duty being charged. At the moment the anomaly persists that an operator using a vehicle of a maximum Gross Plated Weight (e.g. 38,000 kg permissible on 5 axles) on certain types of operation (e.g. using only 4-axle combinations) may be kept to a lower permissible Gross Plated Weight (in this case 32,520 kg) and nevertheless be taxed at the higher (38,000 kg) rate.

Vehicles of over 30 cwts Unladen Weight exempt from Plating and Testing (including Dual Purpose Vehicles) will be taxed at the lower rate in the schedule for vehicles of up to 12 tonnes Gross Plated Weight.

If a vehicle is altered in such a way as to bring it into a higher tax rating, or its use is altered with the same effect, the owner is responsible for notifying the Local Vehicle Licensing Office of the alteration. Further, some alterations may convert a vehicle from a type which was not chargeable with car tax, into a type which is so chargeable (e.g. the fitting of side windows in a van to the rear of the driver's seat). Such alterations must be declared to a local Officer of Customs and Excise.

3.1.3 Fuel Tax and Rebated Fuel Oil

Operators must use diesel fuel on which the full rate of fuel tax has been paid.

It is an offence to use fuel on which the lower rate of duty has been paid ('rebated' fuel), unless its use is for purposes other than driving road vehicles (e.g. for auxiliary equipment, contractor's plant etc.) (15).

3.1.4 Trade Plates

The old distinction between General and Limited Trade Licences has been replaced by one form of Trade Licence (16).

Trade Licences allow Motor Traders (17) who exhibit the Trade Licence Plate (red

---

(14) Finance Act 1982, section 8 - not yet implemented at the time of writing.
(15) Hydrocarbon Oil Duties (Rebates and Relief) Regulations 1964.
(16) Vehicles (Excise) Act 1971, section 16.
(17) Defined as manufacturer, dealer or repairer (including fleet operator). A motor vehicle operator (haulier or private carrier) may use Trade Plates in connection with their own vehicle repair shops, if they can satisfy the local taxation officer that a genuine repair shop is in operation.

numerals and letters on a white background) to use an otherwise unlicensed vehicle in connection with their business as Motor Traders, but not for the carriage of any goods in connection with any other business or for hire or reward. Examples of Trade Use would be:-

(a) Use of a recovery vehicle kept by the trader (18);
(b) Testing or demonstrating a vehicle (19);
(c) Proceeding to a weighbridge to ascertain unladen weight, and returning;
(d) Collection, delivery or removal of a vehicle.

The trader may only use Trade Plates on vehicles which are **temporarily** in his possession, apart from recovery vehicles.

Vehicles may be used with Trade Plates for the carriage of certain loads, including:-

(a) parts, accessories and equipment;
(b) loads needed for demonstration or test purposes, and which are returned to the point of loading (20).

3.2 Driving Licences

This section describes two different types of licence:-

(a) An ordinary driving licence; and
(b) The Heavy Goods Vehicle Driving Licence.

Distinctions can be made between these two types of licence on the basis of the issuing authority, the scope, and the legislation under which each is required. There is provision in the regulations relating to HGV Licences for dual testing, i.e. a person who passes the HGV Drivers' test shall be deemed to have passed the ordinary motor car test for vehicles of a class which his age group allows (21).

The holder of an ordinary provisional driving licence and an HGV provisional driving licence may drive an HGV and take a dual test (21).

3.2.1 Ordinary Licences

(1) Minimum Age for Driving No person under 17 years of age may drive a mechanically propelled vehicle on a road to which the public have access, and no person under 21 years of age may drive a Heavy Goods Vehicle, Tractor or Locomotive. There are two exceptions:-

(a) Persons of 16 years and over may drive mopeds (22);
(b) Articulated vehicles may not be driven by drivers under 21 years of age, unless the tractive unit is under 2 tons Unladen Weight (23).

A medium goods vehicle, i.e. between 3.5 tonnes and 7.5 tonnes permitted maximum laden (or train) weight, may be driven at 18 years of age. A small goods vehicle, i.e. under 3.5 tonnes Gross Plated Weight, may be driven at 17 years of age.

---

(18) Trade Plates cannot be used by a lorry to carry scrap cars (Gibson v. Nutter 1983), since "recovery" relates to "disabled", not scrapped or abandoned, vehicles.
(19) A prospective purchaser may be carried.
(20) Road Vehicles (Registration and Licensing) Regulations 1971.
(21) HGV (Drivers' Licences) Regulations 1977, section 25. The vehicle would have to display both an L-plate and an HGV L-plate. It is doubtful if such a trainee could then drive on a motorway, since he would be driving 'by virtue only of his being the holder of an (ordinary) provisional licence under section 88(2) of the Road Traffic Act 1972'.
(22) Mopeds - mechanically propelled cycles with an engine capacity not exceeding 50 c.c.
(23) Road Traffic (Drivers' Ages and Hours of Work) Act 1976.

(2) Driver's and Employer's Responsibilities It is an offence to drive a mechanically propelled vehicle on a road to which the public have access, unless the driver holds an ordinary driving licence appropriate to the class of vehicle being driven (24). These classes are given on the inside cover of every ordinary driving licence.

A person who steers a vehicle which is being towed is now regarded as a driver (25).

To qualify for a licence, a driver must have held a full driving licence during the last ten years, or have passed a driving test appropriate to the class of vehicle which it is intended to drive. If a driver wishes to drive another class of vehicle he will require to take an additional test. To enable a person to learn to drive, provisional licences are issued by DVLC, Swansea (26). The learner must be accompanied by a full ordinary licence holder, display L-plates, and keep off motorways.

It is an offence to employ an unlicensed driver (27).

The licence must be signed by the licence holder, and must be produced to a police officer on demand, or within five days at a Police Station named by the licence holder.

Ordinary Driving Licences are issued by the Driver and Vehicle Licensing Centre (DVLC), Swansea. They are renewable at age 70.

(3) Disqualifying Offences (28) The Road Traffic Act 1972 greatly increased the penalties for certain traffic offences, and provided that the courts could disqualify from driving for any of the offences in the following list. The Transport Act 1981 introduced the concept of 'penalty points' for these offences. The 'penalty points' which the court may award for each offence are shown in brackets after the name of the offence in the list.

(i) Causing death by reckless or dangerous driving;
(ii) Racing on the highway;
(iii) Driving or attempting to drive whilst under the influence of drink or drugs (29);
(iv) Reckless driving (10 points);
(v) Dangerous driving (10 points);
(vi) Careless or inconsiderate driving (2 - 5 points);
(vii) Driving under age (2 points);
(viii) Being in charge of a motor vehicle whilst unfit through drink or drugs (10 points) (29);
(ix) Failure to comply with a traffic direction (3 points) (30);
(x) Failure to stop after an accident (5 - 9 points);
(xi) Driving without a licence (2 points);
(xii) Using a vehicle without insurance (4 - 8 points);
(xiii) Taking a vehicle without the owner's consent, driving it or allowing oneself to be carried in it (8 points);

---

(24) Road Traffic Act 1972, section 84(1).
(25) McQuaid v. Anderson, 1980.
(26) A first-time applicant for a Provisional Driving Licence may start learning to drive once his application is received at DVLC. (Transport Act 1981).
(27) Road Traffic Act 1972, section 84(2).
(28) Road Traffic Act 1972, sections 93 following, and Schedule 4.
(29) Or with breath/alcohol concentration above the prescribed limit.
(30) Courts have no power to endorse a licence for failure to stop on the direction of a Traffic Warden (as opposed to a Police Constable) directing traffic (Rumbles v. Poole 1980).

(xiv) Driving with defective eyesight, or refusing to submit to an eyesight test (2 points);
(xv) Driving whilst disqualified (6 points);
(xvi) Contravention of Construction & Use Regulations (3 points) (31).

It is mandatory for a court to disqualify for at least one year for any of the above offences (i) to (iii) inclusive. Disqualification for any of the offences (iv) to (xvi) inclusive is discretionary, but endorsement of the licence must be ordered unless there are special reasons for the court to decide otherwise. A person collecting 12 or more penalty points within three years will almost always be automatically disqualified for at least six months, unless the court decides that there are special reasons to disqualify for a shorter period or not at all. Once a period of disqualification has been imposed, the 'slate is wiped clean', and the points are not counted again, but, to discourage further offences, a subsequent disqualification within three years will be for a progressively longer period (32).

Special Reasons In the case of mandatory disqualifications, the only special reasons which the court may consider are reasons relating to the offence itself. However, in the case of endorsement and 'totting up', the special reasons which the court may consider are wider, and include all the circumstances of the case (33), including, it is suggested (34), hardship to the defendant. In the case of offence (xvi) above (contravention of Construction & Use Regulations), special reasons may be found where the defendant can prove that he did not know, nor had he reason to suspect, that he was using or causing a vehicle to be used contrary to the C & U Regulations (35). Such an offender will not be disqualified, nor will he have his licence endorsed (36). An employer may also be liable for permitting or causing certain offences (e.g. Nos. (vii), (xii) and (xv)), and in such circumstances, if convicted, he could have his licence endorsed and be disqualified. It is, however, a good defence to prove that he did not know, nor could he have reason to suspect, that an offence was being committed (37).

Drinking and Driving The original 'breathalyser' provisions of the Road Safety Act 1967 (38) have been strengthened by the Transport Act 1981. These provisions apply equally to drivers of goods vehicles. The statutory limit for breath/alcohol concentration becomes 35 microgrammes of alcohol in 100 millilitres of breath (39).

It is an offence to drive a vehicle with an alcohol level exceeding this, but it is worth noting that it still remains an offence under the Road Traffic Act 1972 to drive under the influence of drink, whatever the level. The 1967 Act relaxed the law relating to being in charge of a vehicle whilst under the influence of drink (or drugs). It provides that a person will not be convicted if he can show

---

(31) The court has discretion to disqualify, and an obligation to endorse the licence, if the offence was committed by using, causing or permitting the use of a vehicle or trailer so as to cause, or be likely to cause, danger; or by being in breach of the regulations relating to brakes, steering or tyres.
(32) Transport Act 1981. Existing endorsements under the old 'totting up' provisions count as 3 points.
(33) Road Traffic Act 1972, section 93(3).
(34) 'Road Traffic Offences', Wilkinson, published by Oyez.
(35) Hart v. R., 1957.
(36) Road Traffic Act 1972, schedule 4.
(37) Kerr v. McNeill, 1945.
(38) Consolidated in the Road Traffic Act 1972 as sections 5-13 inclusive.
(39) The equivalent of 80 milligrammes of alcohol in 100 milliltres of blood, or 107 milligrammes of alcohol in 100 millilitres of urine.

that he had not driven while he was under the influence, and that there was no likelihood of his doing so.

The 'breathalyser' procedure is complicated. A Police Constable may require a person to take a breath test if he has reason to suspect that that person has been drinking, or has committed a moving traffic offence, or been involved in a road accident. If the test is refused, or if it gives a positive result, the suspect can be arrested and taken to a police station. Any conviction will be based upon the results obtained by the very much more sophisticated electronic breath machines used at the police station. A suspect will have the right to request a blood sample test where the breath-analysis (which gives an instant quantifiable reading) does not exceed 50 microgrammes of alcohol per 100 millilitres of breath. Police powers to require a breath test from drivers and persons suspected of having committed a moving traffic offence, are extended to persons who have been driving, or have been in charge of a vehicle, and who are similarly suspected. Police may also enter premises to require a breath test from a person suspected of driving whilst impaired, or who has been driving or in charge of a vehicle involved in an accident in which another person has been injured. The penalties for hit-and-run offences committed to avoid being breathalysed are greatly increased. Failure to provide a specimen incurs the same penalties as the main offence.

Removal of Endorsement and Disqualification There is provision in the Road Traffic Act 1972 for a person disqualified for more than two years to apply for removal of the disqualification after certain periods laid down in the Act (40). There is also provision for a person to obtain a clean licence from the DVLC on surrender of his endorsed licence, at the end of the period of endorsement (41).

3.2.2 Heavy Goods Vehicle Drivers' Licences

It is an offence (42) to drive a Heavy Goods Vehicle on a road to which the public have access, unless the driver holds an HGV Licence entitling him to drive the appropriate class of HGV. The licensing of HGV Drivers is controlled by the Licensing Authority (Traffic Commissioners) of the Traffic Area in which the applicant resides.

(1) Qualifying for an HGV Licence - Age Limits The driver must either have been in the habit of driving an HGV of the appropriate class for a period of six months or more within the 12 months preceding 2nd February 1970, or have passed the HGV Driver's Test for the appropriate class.

Applications for licences have to be on form DLG1 obtainable from Traffic Area Offices or Money Order Post Offices. There is provision on the form for an employer to certify the applicant's previous driving experience. The application must be accompanied by a medical certificate (Form DLG1A) if the Licensing Authority requires this. In practice, by agreement between the different area Licensing Authorities, the medical certificate is required
(a) on first application and
(b) by applicants over 60 years of age.

The applicant must not be suspended or disqualified from holding an HGV or Ordinary licence, or suffer from any disability, or be under 21 (17 for vehicle classes 4 and 4A).

---

(40) Road Traffic Act 1972 section 95. The period is 2 years; or half the period of disqualification if between 4-10 years; or 5 years in other cases.
(41) Road Traffic Act 1972 section 101(7).
(42) Road Traffic Act 1972 section 112. It is also an offence to cause or permit the offence, e.g. by employing an unqualified HGV driver (section 112(2)).

Provisional HGV Drivers' Licences may be issued by the Licensing Authorities for the purpose of enabling the holder to learn to drive an HGV.

Learners must be accompanied by an HGV Licence holder for the appropriate class of vehicle, and must display the statutory HGV L-plate. The licence is in the form of a booklet, to be signed by the holder, and it can be suspended or revoked by the Licensing Authority, who will endorse it and return it at the end of the suspension.

The licence must be produced to a Police Officer on demand, or at a Police Station later, in exactly the same way as an ordinary licence; but in addition it must be produced to an Examiner of the Department of Transport, or in default to such examiners' office as may be specified within 10 days.

(2) The HGV Test The test is broadly similar to a car driver's test, but is longer (approximately 2 hours) and incorporates a manoeuvring test on a driving pad, and a wide range of questions embracing not only the Highway Code but also simple vehicle roadworthiness. The driver should present himself for test:-
(a) with an unladen vehicle;
(b) with a seat in the vehicle for the examiner;
(c) in possession of his ordinary driving licence;
(d) with sufficient fuel for the test;
(e) displaying HGV L-plates.

(3) Classes of HGV Licence The following table is self-explanatory.

Class	Type of HGV covered	Also covers other classes
1, 1A	Articulated vehicle	2, 3, 4
2, 2A	Rigid vehicle with more than four wheels (Twin wheels with areas of contact less than 18 inches apart count as one wheel)	3
3, 3A	Rigid vehicle with four or fewer wheels	-
4, 4A	Articulated vehicle with tractive unit less than 2 tons Unladen Weight (43)	-

A = vehicle fitted with automatic transmission.

In each class a licence to drive a vehicle without automatic transmission entitles the holder to drive a vehicle with automatic transmission. A full licence for a lower class can be used as a provisional licence for a higher class.

(4) Suspension, revocation, disqualification and endorsement Where a Licensing Authority suspends or revokes an HGV Licence, the holder must return the licence to him. Suspensions will be endorsed on the licence when it is returned to the holder after the suspension has expired. There is a special appeal procedure to a Magistrates' Court against suspension or revocation of an HGV Licence (44). It

---

(43) An exemption applies where an articulated combination has a Gross Permitted Weight exceeding 7500 kg, or the unit has an Unladen Weight of 15 cwts or less (Heavy Goods Vehicles (Drivers' Licences) Regulations 1977, section 29 (1)(r).
(44) Road Traffic Act 1972 section 118 (above).

should be noted that the grounds for suspension and revocation of an HGV Licence are mainly concerned with drivers' hours and records, overloading, speed restrictions, and use of unroadworthy or prohibited vehicles (i.e. those with a GV9), and as such are different from the Road Traffic Act offences for which a driver's ordinary licence may be endorsed, suspended or revoked. If the holder of an HGV licence has his ordinary driving licence suspended or revoked, he must notify the Licensing Authority and return to him his HGV licence. It will be returned to the driver on production of an ordinary driving licence at the end of the period of his disqualification from driving (45). However, the Licensing Authority may call the driver to a Public Inquiry, and LAs have powers to disqualify persons from holding HGV Licences, if necessary until they have passed an HGV driving test.

<u>(5) The Young HGV Driver Training Scheme</u> (46) Trainee HGV Drivers' Licences (47) may be issued to drivers between the ages of 18-21 who
(a) hold a 'clean' ordinary driving licence <u>and</u>
(b) have entered into a training agreement with an employer registered with the National Joint Training Committee for Young HGV Drivers.

The Trainee HGV Driver's Licence allows the holder to drive HGVs only
(a) for his employer; or
(b) when attending an approved training establishment.

It can be withdrawn on conviction for more than one endorsable road traffic offence.

Class of Trainee HGV Driving Licence	Minimum Age issued	Test permitted
I (Provisional)	20	At age 21
II	19	After 3 months
III	18	After 3 months
IV	18	At age 18

(45) HGV (Drivers' Licences) Regulations 1975, section 12.
(46) Road Traffic Act 1974; Road Traffic (Drivers' Ages and Hours of Work) Act 1976.
(47) HGV (Drivers' Licences) (Amendment) Regulations 1975.

4 Drivers' Hours and Records

4.1 Relevant Legislation on Drivers' Hours

The Drivers' Hours laws are contained in:-
(a) Part VI of the Transport Act 1968, ss.96 et seq.
(b) The Road Traffic (Drivers' Ages and Hours of Work) Act 1976.
(c) EEC Regulations 543/69 and amendments 514-515/72 and 2828/77 (Consolidated Regulations are printed by HMSO).
(d) The Drivers' Hours (Harmonization with Community Rules) Regulations 1978.
(e) The Community Road Transport Rules (Exemptions) Regulations 1978.
(f) The Drivers' Hours (Passenger and Goods Vehicles) (Modification) Order 1971.
(g) The Drivers' Hours (Goods Vehicles) (Exemptions) Regulations 1978.

With the implementation from 26th January 1978 of EEC Regulation 543/69, substantial changes were made to the statutory rules. The new regulations were fully implemented on 1st January 1981.

There are now three basic sets of rules as follows:-
(a) for drivers on international journeys;
(b) for UK drivers subject to EEC Regulation 543/69 (i.e. performing National work).
(c) for UK domestic drivers exempt from EEC Regulation 543/69 but subject to the hours rules contained in the Transport Act 1968.

In view of the content of EEC Regulation 543/69 (as amended), it has become necessary to differentiate between drivers operating under Community (i.e. National and International work) and Domestic rules, insofar as permitted hours, breaks and rests are concerned; and the Harmonization Regulations 1978 (referred to in full above) were made in consequence.

Complete exemption from laws governing Drivers' Hours only applies to the Armed Forces, the Police and Fire Brigades (1) (other Crown vehicles are <u>not</u> exempt).

4.2 Scope of the Regulations

The regulations apply to the drivers of almost every vehicle constructed or adapted for carrying goods, irrespective of whether or not the vehicle requires an Operator's Licence, or whether or not the driver is required to keep records.

They apply to all domestic journeys or work, and to Community regulated (i.e. National and International) journeys and work.

---

(1) The Drivers' Hours (Passenger and Goods Vehicles) (Modification) Order 1971.

4.3 Journeys and Work

The following definitions are relevant to journeys or work:-

Domestic journeys or work - journeys or work to which the Community rules do not apply but to which the provisions of the Transport Act 1968 do apply (e.g. journeys made by exempted vehicles listed at 4.4.1 below).

International journeys and work - Community-regulated journeys and work which are undertaken in connection with international transport operations, and to which therefore the Community rules apply in full.

National journeys or work - Community-regulated journeys or work which are made in connection with National transport operations in Great Britain, and to which the Community rules apply. Drivers of Goods Vehicles having a Plated Weight exceeding 3.5 tonnes will normally be observing EEC Regulations applying to national journeys or work; but in addition they must comply with the provisions of Part VI of the Transport Act 1968 relating to continuous, daily, and weekly duty periods and 'spreadovers'.

Thus all journeys or work will normally fall into one or more of these categories. The drivers of goods vehicles up to and including 3.5 tonnes Gross Plated Weight are exempt from Community rules, but must comply with the provisions of the Transport Act 1968, section 96. Certain partial exemptions from the Domestic rules apply to vehicles used by doctors, dentists, nurses and vets, and also for some commercial activities. Even in these cases the maximum of 10 hours' daily driving still applies.

4.4 Exemptions from EEC Regulations 543/69

4.4.1 General Exemptions

Drivers of the following vehicles are exempt from the EEC Hours Rules but are automatically subject to the rules of the UK Transport Act 1968, unless also exempt from these as described above.

(a) Goods vehicles and combinations below 3.5 tonnes Gross Plated Weight;
(b) Police, armed forces, civil defence and fire brigade vehicles (1);
(c) Drainage, flood prevention, water, gas and electricity service vehicles;
(d) Highway authority and refuse collection vehicles;
(e) Telegraph and telephone service vehicles;
(f) Radio and TV broadcasting service and detection vehicles;
(g) Post office mail vehicles;
(h) Public authorities' service vehicles so long as they are not in competition with professional hauliers;
(i) Ambulances and rescue vehicles, and other specialised medical vehicles;
(j) Tractors with a maximum speed of under 19 m.p.h. (30 k.p.h.);
(k) Agricultural and forestry machines and tractors;
(l) Circus and funfair transport;
(m) Specialised breakdown vehicles.

4.4.2 Special exemptions from EEC Regulations for vehicles used only on National journeys: In addition to the general exemptions listed above, the following special exemptions apply in the UK only:-

(a) Vehicles undergoing local road test for the purpose of repair and maintenance;
(b) Transport of live animals between farms and local markets and vice versa;
(c) Specialised vehicles used (i) at local markets;

---

(1) The Drivers' Hours (Passenger and Goods Vehicles) (Modification) Order 1971.

(ii) for door-to-door selling;
(iii) for mobile banking;
(iv) for worship;
(v) for the lending of books, records or cassettes;
(vi) for cultural events or mobile exhibitions.

(d) Transport of milk between farms, dairies and distribution centres.

Drivers exempt from EEC Regulation 543/69 by the above derogations will still be subject to the hours rules of the Transport Act 1968 unless exempt from these also.

4.5 Definitions

4.5.1 Driver and Crewman

The EEC Regulations apply to drivers of vehicles with a Gross Plated Weight exceeding 3.5 tonnes when any trailer or semi-trailer is included. The 'Driver' is any person who drives the vehicle, even for a short period, or who is carried in the vehicle in order to be available for driving if necessary.

Some of the rules apply to crew members, i.e. drivers and mates; other rules apply to drivers only. A 'driver's mate' is not intended to mean a person who accompanies a driver to assist with loading and unloading, but is intended to mean the second man who is present because of a statutory requirement, e.g. in the case of an abnormal indivisible load, as an attendant, or when the vehicle is 'double manned', or when the second man accompanies the driver to assist him in the transport operations (2).

UK Regulations define two types of driver:-

(a) an employee driver who drives a vehicle in the course of his employment; and
(b) an owner driver who drives a vehicle for the purpose of his own trade or business.

A person driving both passenger and goods vehicles is considered to be a goods driver when in a working week, at least half of his time is spent in driving, or on work in connection with, goods vehicles, whether the driving and work applies to Community-regulated or Domestic journeys. However, work done with passenger vehicles normally counts towards the build-up of goods vehicle driver's hours.

4.5.2 Driving Time

Time spent at the controls of a vehicle with the engine running, for the purpose of controlling its movement, whether it is moving or stationary (3). This is an important definition which becomes even more precise with the full implementation of provisions relating to the compulsory fitting and use of tachographs in goods vehicles (4).

Continuous driving is defined by the EEC Regulations as time spent at the controls of a vehicle for the purpose of controlling its movement, whether or not it is in motion and with the engine running. Any interruptions from driving, for loading, unloading and other non-driving work, can be completely ignored for the purposes of the continuous driving limit.

---

(2) EEC Regulation 543/69, Article 1.3; also Pickfords Removals v. Metropolitan Deputy Licensing Authority, 1981.
(3) Transport Act 1968, section 103(3).
(4) Transport Act 1968 section 97, and Passenger and Goods Vehicles (Recording Equipment) Regulations 1979.

4.5.3 Working Day and Working Week

The Transport Act 1968 defines both a 'Working Day' and a 'Working Week':-

The Working Day begins at any time of the day or night, and continues until there is a statutory rest period. The driver is held to be on duty during his working day, and off duty during his period of rest (subject to the spreadover provisions in the Transport Act 1968) (5).

Duty time covers any time spent on duty by an employee in employment which involves driving a goods vehicle. Duty is not limited to driving time, or to time spent working or in connection with the vehicle or its load, but includes all such activities, as well as any further time spent under the direction of the employer. However, an employee is not on duty (a) during breaks for rest and refreshment if during the break he has no specific duties or responsibilities to discharge for his employer and (b) so far as a driver is engaged on Domestic journeys and work, if he is carrying out duties for an employer for whom no driving is performed.

It is important to be able to establish what counts as duty and hence impinges on (cuts into) a driver's rest period.

As far as drivers operating under GB Domestic Rules are concerned, some guidance is available from a number of decided cases. Two important rules should certainly be remembered.

In the first place, it is generally accepted that any activity from which the employer gains a benefit will count as duty. For example, in

Potter v. Gorbould 1968 a driver was held to have impinged on his rest period by doing non-driving work for his employer, in this case, cutting up scrap metal; and in

Mitchell v. Abbott 1966 where the driver at the end of his driving period returned to his own home in a relief car, the court held that the time taken to travel home did not count as duty time. He had exercised an option to drive himself home for his own benefit. (He was paid for the time during which he drove home, and his being at home that evening made his scheduling the next day so much easier for his employer!)

Secondly, the fact that payment is made is not conclusive. This is illustrated by Mitchell v. Abbott (1966). It is not unusual for employers to pay drivers a premium payment to cover such inconvenience, for guaranteed days of certain lengths, or for periods off-duty during spread-overs which are 'paid through'; all of which when totalled amounts to more than the permitted hours.

This, however, does not give any evidence of, or constitute, an offence. In **Burridge v. Alderton (Traffic Examiner, Eastern Licensing Authority), 1974**, the court accepted that the test as to whether a driver is on duty or not, is not payment, but simply whether the driver is under the employer's control.

As far as drivers on National journeys and work (but not Domestic work) are concerned, the courts have recently placed some very strict interpretations on the meaning, under EEC rules, of driving and rest (and, by implication, duty). In a magistrates' court in Sussex in 1982, drivers of Goods Vehicles over 3.5 tonnes GPW were successfully prosecuted for not recording, as driving, empty running not in connection with their, or their employer's, business. (They were driving the

---

(5) Transport Act 1968, section 96(3).

vehicles as 'floats' in a village carnival!) In that same year a driver who worked as a non-driver (assembly-line worker) for another employer was held to have impinged on his statutory rest period (i.e. to have been on duty) (Pearson v. Rutterford and Another, 1982).

The Transport Act 1968 (6) specifically rules that an owner-driver is considered to be off duty when not driving a goods vehicle in the course of his trade or business, or doing work in connection with a vehicle or its load for the purpose of his trade or business. However, there is no parallel provision in the EEC Regulations applicable to Community-regulated work. Indeed, the Department of Transport have advised (7) that all driving of Community-regulated (National and International, but not Domestic) work, both in connection with business and for private social purposes, should be recorded as driving time.

The Working Week is defined, in relation to Domestic Work, by section 103 of the Transport Act 1968 as the week beginning midnight Saturday/Sunday, but in the same section provision is made for an operator to apply to the Traffic Commissioners to have his Working Week altered to commence at midnight between any two other days. The Working Week, as it affects a driver engaged on International or National journeys and work, or on mixed National and Domestic journeys and work, means any period of seven consecutive days. It is frequently described as a 'rolling week'.

4.5.4 Breaks

There is a fundamental difference between the 11 hours period of rest and the half-hour break which a driver is required to have after a continuous period of working (if he is performing domestic work) or driving (if he is performing Community-regulated work). This half-hour break is defined as a period during which the driver is able to obtain both rest and refreshment. The break may be on duty, or it may be off duty; rest is by definition off duty. The driver would be taking a break on duty, for example, if he was bound by the terms of his employment to remain on or near his vehicle. Thus, a driver who has a flask and a sandwich in a lay-by would be taking a break on duty; whereas a driver who leaves his vehicle at a branch depot and walks into town to have a meal at a cafe would obviously be taking a break off duty. Between these two extreme examples there will obviously be many difficult cases where it would be hard to decide whether a driver was on or off duty. Nevertheless it is important to try to be definite, because where the working day is spread over more than 11 hours and the driver is performing National or Domestic work, a break off duty will not count towards his aggregate working day.

4.6 Specific Requirements of the Drivers' Hours Regulations

There follows a section-by-section guide to the hours regulations.

4.6.1 Maximum total daily driving time

A driver on Domestic work or journeys must not, in any working day, drive a goods vehicle for periods which amount in aggregate to more than 10 hours (8). Drivers on National and International journeys and work are restricted to 8 hours (9)

---

(6) Transport Act 1968, section 95(3).
(7) 'Headlight', April 1980 - report of cases brought by Sussex Police, and the Department of Transport's interpretation.
(8) Transport Act 1968, section 96(1).
(9) This may be extended on two occasions in the working week to 9 hours, if the driver does not drive a restricted vehicle. Community Drivers' Hours Rules (Temporary Modifications) Regulations 1978; EEC Regulation 543/69 Article 7.

maximum total daily driving time.

4.6.2 Continuous driving without a break

Drivers and crew members on National and International journeys or work must take a break for rest and refreshment of at least 30 minutes after 4 hours (10) continuous driving. A driver may instead, however, take two breaks of not less than 20 minutes each, or three breaks of not less than 15 minutes each, in which case the breaks can be spread out over and immediately following the continuous driving period.

The driver of a non-rigid restricted vehicle (one drawing a trailer or semi-trailer when the permissible maximum weight exceeds 20 tonnes, or one drawing more than one trailer) must take a break of at least one hour. This break can be replaced by at least two breaks of not less than 30 minutes each, spaced over the daily driving period in such a way that no continuous driving of more than 4 hours is performed. The continuous driving limit does not apply to drivers of vehicles operating within 50kms (31 miles) of base (including municipalities the centre of which is situated within that radius), but they must take sufficient breaks from driving to ensure that the total break periods are observed.

The maximum continuous driving without a break which a driver on Domestic journeys or work can drive, is limited only by the maximum amount of continuous duty prescribed by the Transport Act 1968.

4.6.3 Continuous duty without a break

Drivers on Domestic and National journeys and work must take a break for rest and refreshment of at least 30 minutes after they have been continuously on duty for 5 hrs 30 mins (11). Note that the Act says 'on duty', which is not necessarily the same as driving.

For National journeys or work, or mixed Domestic/Community journeys and work, two breaks of 20 minutes or three breaks of 15 minutes may be substituted. When a restricted vehicle is driven, the statutory break is then for one hour or two 30-minute periods. Generally the break will occur before 5 hrs 30 mins continuous duty, i.e. after 4 hours continuous driving.

4.6.4 Maximum total weekly and fortnightly driving and duty time

Drivers on International and National journeys and work must not drive in aggregate for more than 48 hours in a working week, nor for more than 92 hours per fortnight (12). Drivers on National and Domestic journeys and work are restricted to 60 hours total time on duty in any working week (13). There is no limit on weekly driving time for drivers on Domestic journeys and work.

4.6.5 Maximum daily duty/spreadover limits

On National and Domestic journeys or work the maximum length of the driver's working day is 11 hours, but this may be spread over a longer period of up to 12 hrs 30 mins provided that the amount by which the working day exceeds 11 hours (in the case of a 12 hr 30 mins spreadover this would be 1 hr 30 mins) can be taken some time during that working day as a period off duty (statutory off-duty

---

(10) Drivers' Hours (Harmonization with Community Rules) Regulation 1978; EEC Regulation 543/69 Article 8.
(11) Transport Act 1968, section 96(2).
(12) EEC Regulation 543/69 Article 7(4). The week referred to here is the 'rolling week' of any 7 consecutive days.
(13) 0001 hours Sunday to 2359 hours Saturday.

break) (5). It is this which makes it so important to differentiate between breaks on and off duty, as obviously only the latter would be permissible where the working day was spread over more than 11 hours.

Where a vehicle operating on International or National journeys or work is double-manned by two drivers, the following spreadovers are allowed.

Double-manned - No Bunk 17 hours' spreadover, provided at least 6 hours is taken completely off duty, and each driver has a daily rest period of not less than 10 consecutive hours in the 27-hour period preceding any given time when he is on duty.

Double-manned - With Bunk 22 hours' spreadover, provided at least 11 hours is completely off duty, and each driver has a daily rest period of not less than 8 consecutive hours in the 30-hour period preceding any time when the driver is on duty.

The bunk facility must enable the co-driver to lie down comfortably, and the statutory rest period may only be taken on the bunk if the vehicle is stationary; otherwise it must be taken away from and outside the vehicle. However, if the vehicle is double-manned, breaks (but not rest periods) may be taken while the other man is driving.

4.6.6 Daily rest

A driver, or a mate, on International or National journeys or work must have a daily rest period of not less than 11 consecutive hours during the 24-hour period preceding any time when he is driving or doing other work in scope. ALTHOUGH THERE ARE NO SPREADOVER PROVISIONS IN THE E.E.C. REGULATIONS, THE EFFECT OF THE ABOVE IS TO ESTABLISH A MAXIMUM DAILY SPREADOVER OF 13 HOURS (i.e. 24 minus 11 hours).

This period of 11 hours can be reduced to 9 hours twice each week when the rest is taken at the place where the vehicle is based; or to 8 hours twice each week when the rest is taken at a place other than where the vehicle is based. Rest time 'lost' in this way must be compensated on other days throughout the week (14).

For drivers on Domestic work or journeys the Transport Act 1968 (15) prescribes an interval of rest of at least 11 hours between each working day. It thus implicitly defines a working day as a period of duty separating two periods of rest (not breaks off duty) of at least 11 hours.

4.6.7 Weekly rest

Drivers, and mates, on International and National journeys and work must take a weekly rest period of at least 29 hours immediately preceded or followed by their daily rest period. Of this 29 hours, not less than 24 hours must be consecutive and preceded or followed by a daily rest period; the remaining 5 hours may be taken at any other time during the working week in compensation (16).

Within every working week a driver on Domestic journeys or work must have a period of rest of at least 24 hours (17). It should be noted that:-

---

(14) EEC Regulation 543/69, Article 11.
(15) Transport Act 1968, section 96(5).
(16) EEC Regulation 543/69, Article 12. The week is a 'rolling week' of any 7 consecutive days.
(17) Transport Act 1968, section 96(5). The week is a fixed week of 7 days from 0001 hrs on Sunday.

(a) this period does not have to run from midnight to midnight;
(b) the 24 hours rest period can run into the next following week, provided that the amount by which it does so is not counted again as rest in the second week.

4.7 Miscellaneous Conditions

4.7.1 Trains and Ferry Boats

When a driver engaged in the carriage of goods accompanies a vehicle transported by train or ferry-boat on International or National journeys, the daily rest period may be taken partly on the train or ferry and partly on land, interrupted not more than once, providing:-

(a) the interruption to the rest period does not exceed one hour to include customs facilities etc;
(b) the rest taken on boat or train must be preceded or followed by rest taken on land;
(c) during both portions of the rest period, the crew members must have access to a bunk or couchette;
(d) a rest period so interrupted must be extended by two hours to compensate for the interruption.

Note that any time on board which does not as above count towards daily rest, may be regarded as a break off duty.

4.7.2 Emergencies

The Transport Act 1968 (relating to Domestic journeys and work and, so far as duty and spreadover provisions are concerned, to National journeys and work) and the EEC Regulation 543/69 (relating to International and National journeys and work so far as driving time, rests and breaks are concerned) provide different interpretations of what constitutes an 'emergency' under which drivers may exceed the limits on driving, rest and duty. The relevant definitions and exemptions are given below. Note that by and large the EEC definition of 'emergency' embraces the UK definitions of 'emergencies and unavoidable delay'.

4.7.2.1 Emergencies and unavoidable delays (UK Domestic Regulations)

An emergency is 'an event requiring immediate action to prevent danger to life or health, serious disruption of essential public services (gas, water, electricity or drainage) or of telecommunication or postal services, or disruption of the use of roads and airports, or damage to property'.

In such an emergency a driver need not observe the ten hours' driving rule, the half-hour 'break' rule, the 'spreadover' regulations (5) or the 60 hour working week (18). He need not take the statutory day off in the working week if the previous 'day off' (24 hours) was interrupted by the emergency, provided that he takes his 'day off' in the next following three weeks.

A driver will not be convicted for contravening the regulations if he can show that he was unavoidably delayed by unforeseen circumstances beyond his control (19).

A driver may continue working and driving in order to deal with an emergency, but he may still not exceed the period of 11 hours on duty unless the emergency continues beyond the 11 hours, after which a driver cannot resume duty without first taking 10 hours' rest (or an aggregate of 10 hours if the rest is

---

(18) Transport Act 1968, section 96(5).
(19) Transport Act 1968, section 96(11).

interrupted by the emergency).

4.7.2.2 Emergencies (International and National journeys and work)

In the event of an emergency on the road, the driver can break the daily and weekly driving periods and the daily rest requirement to the extent necessary to ensure the safety of people, the vehicle and its load. He must note the reasons for doing so on his tachograph disc. On National journeys and work (only), if the emergency falls within the scope of the definition in the UK Domestic Regulations (above), the driver is also exempt from the National limits applied to daily duty, daily spreadover, continuous duty, breaks from duty, and weekly duty; but there is no exemption concerning the extension of any continuous driving period beyond 4 hours.

4.7.3 Part Time Drivers

On National and Domestic journeys and work, a driver who drives for less than 4 hous on each day of the working week is exempt from the following requirements of the Transport Act 1968 (20):-
section 96(1) - Aggregate driving time will obviously not apply;
section 96(2) - The half-hour break rule will equally not apply;
section 96(3) - The working day will not be limited;
section 96(4) - The 11-hour rest rule will not be applied;
section 96(5) - The working week will not be limited;
section 96(6) - There will be no requirement to have 24 hours' rest.

However, if on any day of that working week the driver drives for more than 4 hours, all the above provisions must be observed for that week, even reckoning retrospectively.

In computing the length of the working week, which is really the sum of the lengths of the working days in the working week, the length of a working day on which the driver does not drive at all (but does work on non-driving duties) may, if this exceeds 11 hours, be counted only as 11 hours (21). It is necessary, however, for the driver to take an 11 hour interval of rest each working day.

International Journeys and Work The EEC regulations make no provisions for part-time drivers. Unless otherwise exempt, a part-time driver is fully subject to the regulations, no matter how short the driving period.

4.7.4 Agriculture, Forestry and Building etc Time spent elsewhere than on a road in the course of operations of agriculture, forestry, building, civil engineering, construction and quarrying is not counted as driving time for the purposes of calculating the amount of driving undertaken in a day (22). However, it will count as duty time if time spent driving on public roads exceeds the 4 hours limit for part-time drivers (above).

4.7.5 Special Needs

National and Domestic journeys and work - The Drivers' Hours (Goods Vehicles) (Modification) Order 1970 exempts from the hours and rest restrictions drivers of small goods vehicles (23) which are used for professional purposes by doctors, dentists, nurses, midwives, vets, commercial travellers, AA and RAC scouts, and

---

(20) Transport Act 1968, section 96(7).
(21) Transport Act 1968, section 96(8).
(22) Transport Act 1968, section 96(9), as extended by the Drivers' Hours (Passenger and Goods Vehicles) Modification Order 1971.
(23) i.e. under 3.5 tonnes Gross Plated Weight.

fitters on repair and maintenance duties, with the exception of the 10 hour driving rule which all the above must still observe.

The Drivers' Hours (Goods Vehicles) (Exemptions) Regulations 1978 relax the hours and rest regulations by making concessions to drivers engaged on certain activities; a schedule to the regulations lists the activities concerned and the extent to which, in each case, sections 96(3), 96(4), 96(5) and 96(6) of the Transport Act 1968 are relaxed.

National and International Journeys - Transport of Harvest Produce Under EEC rules the normal daily rest period is reduced to 10 hours provided the reduction is made up by an equal amount by addition immediately before or after the weekly rest period. The driver may only take advantage of this concession for one harvest period, and on not more than 30 days per year.

4.7.6 Bonus Schemes - International and National Journeys

Bonus or incentive schemes related to distance travelled and/or tonnage of goods carried are not negotiable unless they are of such a kind that they are unlikely to constitute a danger to road safety.

4.7.7 Offences, Penalties and Valid Defences

It is an offence to contravene the regulations concerning drivers' hours and rest periods. Contravention makes the driver and his employer liable to a maximum fine of £200 (19).

The Transport Act 1968 introduces two defences in any prosecution for hours offences.

(a) The driver may show that the contravention was due to an unavoidable delay in the completion of a journey. (It is essential that such delays are entered in the driver's record book or on his tachograph disc at the time, to avoid possible future misunderstandings).

(b) An employer or transport manager can show that he did not cause or permit the contravention, if he can show that it arose out of the driver working for another employer, and that he - the employer or transport manager - could not reasonably have been aware of this.

4.7.8 Mixed UK Domestic and Community Driving

When a driver is engaged on mixed domestic and Community-regulated work, he may elect to abide by the least onerous criteria. Thus, provided he does not exceed the Community limit of 8 hours' driving on the Community-regulated part of his work, he can drive up to the Domestic level of 10 hours per day.

However, where the Transport Act 1968 is more restrictive, for example in the matter of the daily and weekly limit on 'work', he must observe this requirement regardless of the split between domestic and Community-regulated journeys and work. Obviously, he will always be within the law if he simply observes the National journeys and work regulations, which are the product of the harmonization of the two sets of rules. Where an International journey is performed during the period in question, he must observe the International journeys and work provisions for the whole of that journey.

(19) Transport Act 1968, section 96(11).

4.8 Drivers' Records

4.8.1 Relevant Legislation

The law relating to the Records of Work of Drivers of Goods and Passenger Vehicles is contained in:-

(a) The Transport Act 1968, Part VI, Sections 96ff;
(b) EEC Regulations 543/69 and 1463/70;
(c) The Drivers' Hours (Keeping of Records) Regulations 1976;
(d) Passenger and Goods Vehicles (Recording Equipment) Regulations 1979;
(e) The Community Road Transport Rules (Exemptions) (Amendments) Regulation 1980.

4.8.2 Scope of the Legislation

All drivers and crew members of vehicles subject to the Community Hours Regulations are bound by Regulations (24) made under the 1968 Act and the EEC Regulations (above) to keep records of their hours of driving, work and rest. In general, therefore, the drivers and crew members of any goods or passenger vehicle **not** covered by exemption from the Community Regulations - given generally for all International and National work and journeys, or specifically relating to National work and journeys (25), are within scope of the legislation, and in particular drivers of **Goods Vehicles subject to O-licensing which are over 3.5 tonnes Gross Plated Weight, and including tractors, trailers and semi-trailers.**

Exemptions This means in fact that drivers of small goods vehicles are not required to keep records; it does not mean that they are not required to observe the hours and rest laws. Where a driver drives both an exempt vehicle and a vehicle covered by the Keeping of Records Regulations, he must keep a record showing the work done on both vehicles during that working day. The Regulations (24), as well as setting out the form of the records, give an important relaxation to some short period drivers who drive a vehicle on Domestic journeys and work on the road for less than four hours in a working day. Such drivers, provided they remain within 25 miles of their Operating Centre, will not be required to keep records for that working day. However, this relaxation is now of very limited use, since most drivers on Domestic journeys and work, **excluding those listed at 4.4.2 and driving vehicles over 3.5 tonnes GPW**, are not required to keep records. It is important not to confuse this provision with the **hours and rest** relaxation for short period drivers, which only applies if the driver drives for less than four hours on each day of a working week, and does not specify any mileage of operation.

4.8.3 Types of Record

The Type of Record used depends upon the kind of work carried out. There are two possible types of record:-

4.8.3.1 Individual Control Book (I.C.B.)

The Community control book is the same as the model International record book already in use under the Drivers' Hours (Keeping of Records) Regulations 1976. It conforms to the model annexed to the AETR Agreement and the EEC Regulation 543/69. It is used for those Domestic journeys and work still subject to the requirement to keep records. Its use, until May 1983, was compulsory on all Non-Community International Journeys, i.e. journeys to a country not in the EEC but

---

(24) Drivers' Hours (Keeping of Records) Regulations 1976.
(25) See above, sections 4.4.1 and 4.4.2, for complete list of general and national exemptions.

to which the AETR Agreement applies. Since May 1983 a tachograph chart can be used as an alternative.

4.8.3.2 The Tachograph Disc

The Tachograph Disc is used on National journeys and work, and its use is compulsory in place of other types of records in other member states of the EEC.

4.8.4 Rules concerning the keeping of records

Every driver required as above to keep records must carry with him a Driver's Record (either an I.C.B. or a Tachograph Disc) which has been issued to him by the user of the vehicle.

4.8.4.1 The Individual Control Book

Form which the records are required to take The statutory record book must conform to the model set out in the regulations (24). It must contain daily sheets (weekly sheets are not allowed). If it contains more than 15 daily sheets, it must contain duplicates with a sheet of carbon or other means of copying. The book must be marked with its serial number, either by perforation or stamping (printing).

The book must not be smaller than A6 size (105 x 148 mm), and comprises:-

On the front cover
(a) Operator's Licence Number;
(b) Serial Number of Book (stamped, printed or perforated);
(c) Name and address of person (employer or owner driver) issuing the book;
(d) Date and country of issue;
(e) Driver's full name, address and date of birth.

Inside the cover
Detailed instructions on completion, with an example.

The Daily Sheets are split into quarter-hour grids covering a 24-hour day, and are completed by drawing a horizontal line across the grids adjacent to the appropriate symbol in the left-hand column. Daily mileage is also recorded. There is a Weekly Summary Sheet using the same symbols. There should be enough space on the front sheet to enter the name, address etc. of a second employer if necessary. There is generally no objection to the inclusion, for instance, on the daily sheet of a limited amount of additional operational material which does not change the character of the record book.

Issue of Record Books The employer is responsible for the issue of record books to each of his drivers. The employer has to ensure that, before the book is issued, the Operator's Licence number is entered on the front cover and each of the sheets, that the book is serially numbered and indicates the number of sheets it contains. A separate serial number must be used for each book issued.

An employer must also issue an I.C.B. (26) to all statutory attendants, as they

---

(24) Drivers' Hours (Keeping of Records) Regulations 1976.
(26) The offence of failing to issue was established in Concorde Transport v. Metropolitan Licensing Authority 1980, and presumably also applies to tachograph discs. It has been established (Lakenby v. Brown of Wem Ltd 1980) that the book supplied to an employee driver need not be an **unused** book.

are held to be crewmen under Article 14 of EEC Regulation 543/69 (27). An operator employing an agency driver must issue the driver with a record (28), as he "employs" (29) the driver by simply engaging him to drive the vehicle, even though there is no contract of employment. The employer must "cause the driver to keep records" (30).

Two or More Employers: Where a driver works for two or more companies, the books must be issued by the company employing him first, but the responsibility always rests with the goods vehicle operator in those cases where a driver works for both a goods vehicle operator and a coach operator.

Where there are two employers, the second employer must enter his name and address on the front sheet of the book. An employer may if he so wishes issue more than one record book to a driver, but entries should not be made in a new book whilst there is space in an old one.

Supply of Information by Employers Each of the driver's employers must, when requested to do so, provide information to the other employers about the number of hours worked by the driver during the preceding 14 days, and details of his last weekly rest period. When a driver leaves his employment, there is a similar obligation on his former employer to provide the information if requested to the driver or his new employer. If the information is requested in writing, it must be supplied in this way (31).

Employer's Responsibilities

(a) Checking Daily Sheets The regulations require an employer to scrutinize and certify each record sheet within 7 days of receiving it from the driver. In the case of single copy record books, each sheet which has been used must be examined by the employer within 7 days of the completed book being handed in, unless in any case the sheet has already been examined and signed when the book was in the employer's possession for checking.

(b) Weekly Sheets The employer is required to check these against the daily sheets and sign in the appropriate space each week. (He may sign **either** the **weekly report sheet** remaining in the book, **or** the copy handed to him and filed).

It is a good defence for an employer to show that he complied with the above requirements as soon as it was reasonably practical for him to do so (32).

(c) Retention Record books, daily sheets and weekly reports, which are returned to the employer who issued them, must be retained by him for a period of 12 months.

(d) Employer's Register of I.C.B.'s Employers must keep registers of record books issued to their drivers. The registers must show the issue of each I.C.B., place of issue, date of issue and subsequent return, and the employer's name and

---

(27) Pickfords Removals v. Metropolitan Deputy Licensing Authority, 1981.
(28) Alcock v. G C Criston Ltd, High Court, 1981. ('Motor Transport' 4th November 1981).
(29) Transport Act 1968, section 103: "employer" (in relation to an employee driver) means the employer of the driver by virtue of whose employment the driver is an employee driver.
(30) Drivers' Hours (Keeping of Records) Regulations 1976, section 14. Under EEC Regulation 543/69, article 14, crew members must keep a record and the 'undertaking' must issue this.
(31) Drivers' Hours (Keeping of Records) Regulations 1976, section 9.
(32) Ibid., section 8(4).

address, and be signed by him. Further information may be added should the employer so wish.

Separate registers may be kept for each depot, or a central register may be used. Registers must be kept for 12 months from the latest of the following dates:-

(a) The date on which the last of the books recorded in the register was returned; or

(b) Where a book entered in the register has not been returned, the date on which the reason for this was entered in the register.

Driver's Responsibilities Drivers must have their current record book with them on their vehicle at all times whilst on duty. Entries in record books and registers must be made progressively (and not retrospectively) as soon as the required information is available.

Erasures are not permissible. All entries should be in ink, ballpoint or indelible pencil. Mistakes can be corrected in the 'Remarks' space. Sheets may not be destroyed (33). In the event of the completion of a journey being delayed for unforeseen circumstances, drivers and their employers must make a note of the circumstances on the sheet containing entries for the day concerned, and initial it.

(a) Return of Completed Record Sheets If the book contains duplicate daily sheets, the driver must hand or send the copy of each completed sheet to his employer within 7 days. If this is not practicable, the sheets must be returned as soon as possible.

Before the sheet is returned, it must first of all be signed by the driver, and the signature entered on the copy at the same time. In the case of a single copy log book, the sheets should be kept in the book until the book is completed.

(b) Returning the Record Book A completed book must be retained by the driver for two weeks after completion, and then handed to the employer who issued it. A book is considered to be completed:-

(a) In the case of a book containing duplicate sheets, when all the sheets have been used;

(b) In the case of a single copy book, when all the sheets have been used, or at the end of 28 days from the date when the first sheet was used, whichever was the earlier. A 'Short Life Book' (15 sheets) can be retained for the two weeks after completion before handing it in, or the employer may call in the book weekly for checking and signing.

If a driver leaves the employment of an employer who issued his log book, then he must return to that employer any log books issued to him, together with all duplicates, unused sheets and weekly reports, and obtain a new log book from his new employer.

(c) Entries by Drivers Detailed instructions are given on the inside front cover of the log books, and there is also a specimen completed sheet in the book. Every detail required on the log sheet must be completed.

The following is a summary of the 'musts' for the driver:-

(a) Check item V (Surname etc.) on front of sheet.
(b) Complete item III (Date book first issued).
(c) Make each entry on the daily sheet as the day proceeds, by means of a continuous line on the chart, showing rest periods, breaks, driving and other work.
(d) Complete weekly report at the end of every week in which one or more daily

---

(33) Sheets are numbered consecutively.

sheets have been made out, and hand one copy to the employer.

(e) Ensure that the means of reproduction is used on duplicate sheets if these are provided.

(f) To complete Boxes 12 to 14 of the daily sheet, the driver must add up the total time spent on each activity as recorded by horizontal lines across the 24-hour graph. In box 12, he must show the total daily rest period before duty first starts, and this will usually include some time shown on the previous sheet. The weekly rest period, and other periods such as holidays, need not be included.

(g) Where stop/start work is being carried out, the driver will not be able to show this accurately on the record. He should simply draw the line to the nearest 15-minute ruling.

(h) The sheet is designed for a 24-hour period starting at midnight. Where, for example, the driver is on night shift, each shift may necessitate the use of two sheets.

(i) If two or more vehicles are driven in the course of a day, this must be clearly shown in boxes 2, 10a, 11 and 13.

The different types of duty and rest are denoted by symbols as follows:-

(a) Daily rest period (bed symbol);
(b) Breaks for rest and refreshment (chair symbol);
(c) Driving time (steering wheel symbol);
(d) Attendance at work (rectangle with diagonal line).

Owner Drivers' Records

If the employer is an owner-driver, and makes a journey in a goods vehicle other than for business purposes, or if he is running empty in the course of his business, he must record the journey (34). The requirement for an owner-driver is the same as that for an employee driver, as far as concerns the completion of an individual control book. The owner-driver must also meet the same obligations in completing tachograph charts where these are applicable. Exemptions from record-keeping are applied to the same circumstances as for employee drivers.

An owner-driver must maintain appropriate registers for recording the issue of control books or simplified record books. He is also required to retain documents of all kinds for the same statutory period as an employer.

Production and Examination of Records

Drivers' record books and employers' registers may be required to be produced for inspection, or called in for inspection, by an officer of the Traffic Commissioner or a police constable. Officers of the Traffic Commissioners include the Certifying Officers, Public Service Vehicle Examiners, and Driving and Traffic Examiners. Such a person may copy any item and any other relevant documents including, for example, wage sheets, and may retain the record if they have reason to believe that a false statement has been made. Examiners or Police may enter premises for this purpose at any reasonable time.

4.8.4.2 Tachographs

Exempt Vehicles

Apart from goods vehicles under 3.5 tonnes GPW being exempt from EEC Hours and Records (Tachograph) Regulations, other vehicles are also exempt from the need to fit tachographs. These fall into two categories:-

(a) Those vehicles, such as municipal refuse vehicles, specialised breakdown

(34) See above, footnote 7.

vehicles etc., to which we have referred under the section on Hours as being covered by UK Domestic Regulations. Drivers of such vehicles do not even need to keep records.

(b) Those vehicles, such as certain specialised vehicles, which are not exempt from the requirement to keep records, even though they are exempt from the EEC Hours and Records Regulations 543/69 and 1463/70. Drivers of such vehicles must keep a I.C.B.

Specific Requirements of the Tachograph Regulations

The Regulations define 'recording equipment' as equipment installed in a road vehicle to show and record automatically, or semi-automatically, details of the movement of the vehicle and certain working periods of its crew.

A tachograph is such a recording device frequently mounted in conjunction with the vehicle's speedometer and capable of giving the following information:-

(a) Speed trace, showing the speed at which the vehicle is driven at any time during the day;
(b) The distance travelled by the vehicle;
(c) A trace showing when the engine is running;
(d) On the EEC type tachograph, an indicator set by the driver to show time spent by the driver driving, on duty and resting.

The instrument is driven either by the speedometer or electrically from a pick-up in the transmission mechanism. It makes recordings with three styluses on a circular chart which is rotated once every 24 hours by a quartz clock mechanism.

If the vehicle is stationary, the three styluses make concentric circular traces. When it moves, two of the styluses automatically record speed and distance. The third stylus is controlled by the 'mode key' which the driver moves to different positions:-

(a) Driving (steering wheel symbol);
(b) Other periods of work (rectangle with diagonal line symbol);
(c) Rest or break (bed symbol) (on a tachograph chart the bed symbol denotes both rest and breaks).

When the activity switch is placed in the 'driving' position (steering wheel symbol), provided the vehicle is moving, the recordings made will appear as a thick black line in the 'driving' field of the chart. If the vehicle is stationary, even if the engine is running, recordings will appear as a thin black line.

With the switch in the other positions, recordings will be made either in the 'rest' (bed symbol) or 'passive work' (crossed rectangle symbol) fields as a thin black line. The 'active work' switch position (crossed hammers symbol) is not used in the UK.

Installation and Inspection of Tachographs

Tachographs are calibrated on being fitted to the vehicle, and sealed to prevent driver interference. They cannot be used under the regulation unless installed, calibrated and sealed by approved fitters in approved workshops (35), using a registered mark on the seals. An installation plaque must be affixed when the tachograph is fitted showing:-

(a) Mark of approved fitter/workshop;
(b) 'w'.....revolutions per kilometre;
(c) 'l'.....metres (tyre circumference);

---

(35) Passenger and Goods Vehicles (Recording Equipment) Regulations 1979 give the Secretary of State for Transport power to grant such approvals.

(d) Date of calibration and sealing of tachograph and plaque.

Tachographs must be inspected every 2 years and re-calibrated every 6 years or after any repair likely to affect their reading.

The following tolerances, plus or minus the actual reading, are allowed:-

	In Use	On installation/calibration
Distance Speed	4% over 1 km. 6 km/hour	1% 3 km/hour
Time	2 minutes per day, or 7 minutes per week	

Use of the Tachograph

Where a tachograph is fitted its use in EEC member states is compulsory. Any other form of record is now only permitted where a vehicle is exempt from the requirements to fit a tachograph.

Employer's Responsibilities

Employers must ensure that sufficient charts are supplied to employees. Employers should bear in mind that these are personal to the owner, and also consider the possibility of the charts being damaged or taken by an approved examiner. Owner drivers must also carry an adequate supply.

The employer must collect completed charts not later than 21 days after use and certify them within another 7 days. Completed charts must be retained for a further 12 months.

Drivers' Responsibilities

The driver must insert a disc in the tachograph every day when starting work, and whenever a change of vehicle takes place. He must write the following information on the chart:-
(a) his name;
(b) date and place chart begins, and the relevant odometer reading;
(c) date and place chart ends, and the relevant odometer reading;
(d) registration number of vehicle(s) used, and the odometer readings relevant to any vehicle changeover.

Each record chart is used only to record personal work of the driver to whom it is issued, and he must transfer it to other vehicles which he drives during the 24-hour period.

The driver must ensure that the instrument is kept reading (at British Standard Time throughout an international journey) and running (with two charts inserted if double-manning), and transfer his chart to any changeover vehicle, entering in the centre field the registration number and odometer reading of the changeover vehicle.

If a tachograph becomes defective during a journey, the chart must be completed by hand or continued on a temporary chart and attached to the original. The equipment itself must be repaired upon return to base or en route, if return does not occur within 7 days (36).

Used charts must be retained by the driver for at least 2 days when operating within the UK, and for at least 7 days when operating on an International journey (37).

The employer must collect completed charts not later than 21 days after use, and certify them within another 7 days, and completed charts must be retained for a further 12 months.

Since May 1983, the replacement of Individual Control Books by tachograph records has been permitted when a driver operates under AETR rules. The use of a tachograph in place of other kinds of records is compulsory in other EEC Member States.

In the case of damage to the sheet, crew members should attach the damaged sheet to a spare replacement. When they are away from the vehicle and unable to operate the equipment themselves, the various periods of time must be entered legibly on the sheet without damaging it, either manually or by automatic recording or otherwise.

Enforcement

Tachographs have to be so designed that it is possible for an authorised examiner to verify, without opening the case, that recordings are being made; and, after opening the case, to read the previous 9 hours' recordings without permanently deforming, damaging or spoiling the sheet.

A driver must be able to produce to an authorised examiner at any time during the period of his statutory retention of his last charts (2 days on National journeys) each chart used in his tachograph.

Examiners may enter the vehicle, inspect the instrument, and inspect and copy or remove any charts found therein. They may also, at any reasonable time, enter premises in which vehicles and records are kept, to inspect them. Any records suspected of being false may be removed and can be held for up to six months.

The rules relating to more than one employer are strict, requiring the driver to notify each employer of the name(s) and address(es) of the other(s) and to return completed charts to the first employer. It is immaterial who issued these.

4.9 Drivers' Hours Records Offences, Defences and Penalties

Any person contravening the requirements of the regulations is liable on summary conviction to a fine not exceeding £200.

It is, however, a good defence for an employer if he can prove to the court that:-

(a) he has given proper instructions to his employees with respect to the keeping of records, and that

(b) he has from time to time taken reasonable steps to ensure that these instructions have been carried out.

Reliance on this statutory defence demands that records should be subject to

(36) EEC Regulation 1463/70, Article 18(1)(2).

(37) EEC Regulation 1463/70, Article 17(5); Article 17(6) allows a 2-day derogation for National journeys in member states.

detailed regular checking, and requires evidence that action has been taken in relation to faults found. Note the defence (section 4.8.4.1 under 'Employer's Responsibilities' above) where Record Books and Sheets have been returned late (38).

Similarly, other defences are provided as follows:-

(a) Tachographs not working:-
  (i) Vehicle proceeding to tachograph centre, or
  (ii) Not reasonably practical for the instrument to be repaired immediately by an approved fitter/workshop (operators are advised to carry in the vehicle a written appointment for the repair), or
  (iii) Manual recordings were being made by driver/crew.

(b) Broken or removed seals:-
  (i) Breakage was unavoidable, or
  (ii) As (ii) above, or
  (iii) The instrument, in all other respects, was being used lawfully.

Traffic Commissioners can revoke, suspend, curtail or prematurely terminate an Operator's Licence in respect of records offences, and a driver's Vocational Driving Licence may be revoked or suspended for similar reasons.

It is an indictable offence to make, or cause to be made, a false entry or an alteration to an entry, with intent to deceive (39).

4.10 Conclusions

The Hours and Records legislation is complicated, but it is important that drivers and employers co-operate in seeing that it is adhered to. The consequences of a detected infringement may not stop with a conviction and penalty imposed by magistrates, but may involve the temporary or permanent loss of some or all of the operator's vehicle O-licences. The importance of an owner-driver's compliance with the regulations is equally obvious. Finally, it should not be assumed that in cases where records are not required to be kept, it is not necessary to comply with the Domestic drivers' hours and work regulations.

---

(38) Drivers' Hours (Keeping of Records) Regulations 1976, section 8(4).
(39) Transport Act 1968, section 99.

FIG. 1 TACHOGRAPH

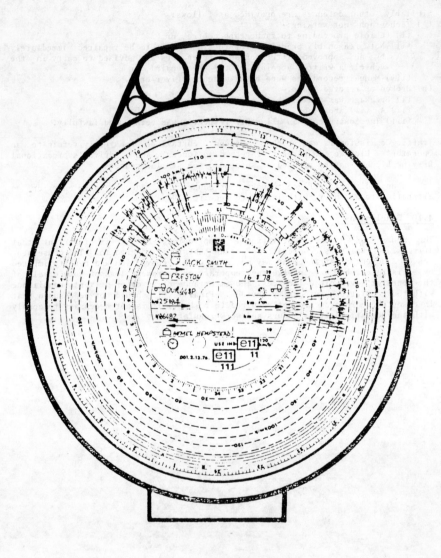

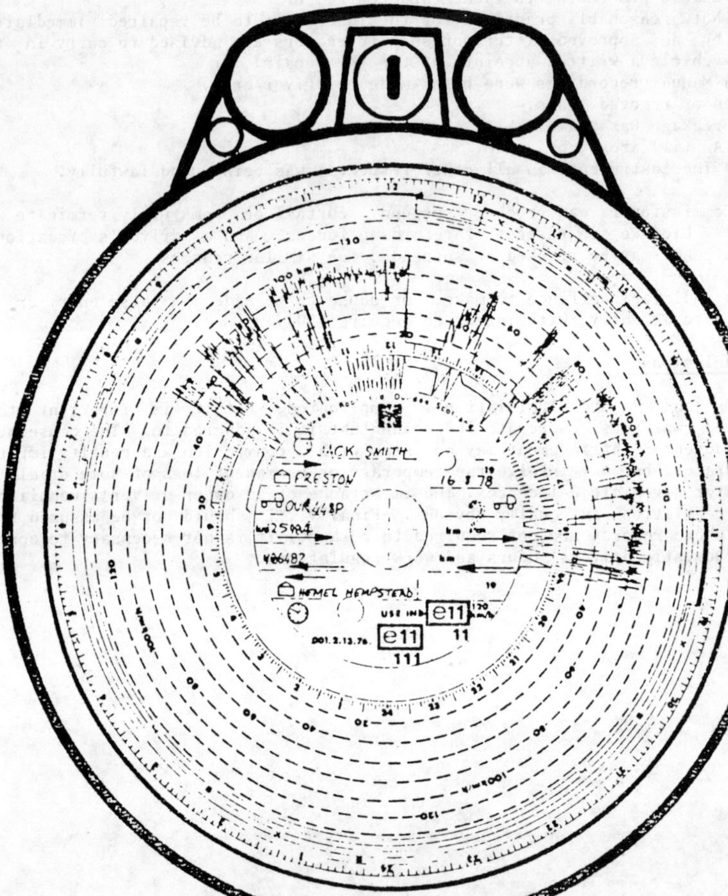

5 Road Traffic Acts and Regulations

5.1 General Traffic Regulations

The Road Traffic Acts (1), as their title implies, concern themselves with enactments intended to promote the safe operation of road traffic. They are also 'enabling' Acts, giving the Secretary of State powers to make regulations about such matters as the **Construction and Use** of vehicles, their **Plating and Testing** and **Type Approval**, the **Licensing** of their drivers, and their **Insurance**. In particular, the **Use** part of the Construction and Use Regulations deals with all manner of traffic matters.

5.1.1 Speed Limits

Speed limits may be of two types - general and particular. The former (general) speed limits refer to all vehicles. They can be looked upon as road speed limits, as they apply generally to all vehicles on a specific road, although different roads will, of course, be subject to different speed limits (2). The latter (particular) speed limits apply to particular vehicles depending upon where they are being driven, although where the general and particular speed limits are not the same in any particular set of circumstances, the lower of the two usually applies (except on Motorways)(3).

If a particular vehicle may be defined in more than one way, and hence would appear to be limited by more than one particular speed limit, the lower of the two will apply (4).

General Speed Limits There is now a general overall maximum speed limit of 70 m.p.h. This applies on Motorways, and on 'derestricted' dual carriageways. On single carriageway 'derestricted' roads, traffic is limited to 60 m.p.h. (5)

On restricted roads, a speed limit of 30 m.p.h. exists under the Road Traffic Regulation Act 1967 (6), but higher or lower speeds may be permitted by order. Earlier Acts used the familiar term 'built-up area', but a restricted road has

---

(1) Road Traffic Acts, 1972 and 1974.
(2) Road Traffic Regulation Act 1967, sections 75 ff.
(3) Road Traffic Regulation Act 1967, schedule 5(13); Motorway Traffic (Speed Limit) Regulation 1974 (S.I.502/1974).
(4) Road Traffic Regulation Act 1967, schedule 5.
(5) Unless either (a) lower 'general' speed limits are in force or (b) the particular class of vehicle is already restricted to lower limits. Temporary Speed Limit Order 1977; Temporary Speed Limits (Continuation) Order 1978.
(6) Road Traffic Regulation Act 1967, section 71(2).

since been redefined (7) as a road provided with a system of street lighting with lamps not more than 200 yards apart, or a road subject to a direction that it shall be a restricted road. In this latter case, in the absence of street lighting, a person cannot be convicted of speeding unless there are the necessary repeater restriction signs, at intervals of not more than 300 yards, or 325 yards where a speed limit of more than 30 m.p.h. is in force.

Certain roads, while not being restricted within the meaning of the Road Traffic Regulation Act 1967 and the Transport Act 1982 (which redefined the term 'restricted road' as above), may nevertheless be subject to a general speed limit. For example, a Trunk Road is not restricted merely because it has street lighting as specified above, unless this was provided before 1st July 1957 (7). Where such a system of lighting exists, but the road is nevertheless not 'restricted' within the meaning of the Acts, evidence of the absence of any derestriction signs will be sufficient to prove the existence of a speed limit. Sections of restricted road furnished with lighting can be specified as unrestricted, and vice versa (8).

Particular Speed Limits Certain classes of vehicle are subject to particular speed limits, either when used on 'derestricted' roads or, in certain cases, also when used on a Motorway (9).

Class of Vehicle	Maximum Permitted Speed when used on		
	Dual Carriageways	Other Roads	Motorways
Passenger vehicles under 3050kg ULW (i.e. Motor Cars), car-derived vans (10) and dual purpose vehicles	70 mph	60 mph	70 mph
Ditto, towing a trailer	50 mph	50 mph	50 mph
Goods Vehicle not over 7500 kg GPW	60 mph	50 mph	70 mph
Articulated Vehicle not over 7500 kg GPW	50 mph	50 mph	60 mph*
Drawbar trailer outfit not over 7500 kg GPW	50 mph	50 mph	60 mph*
Goods Vehicle over 7500 kg GPW	50 mph	40 mph	60 mph*
Articulated Vehicle over 7500 kg GPW	50 mph	40 mph	60 mph*

(7) Road Traffic Regulation Act 1967, section 72. Sub-section (1), as amended by section 61 of the Transport Act 1982, provides for the re-definition of 'restricted roads'; sub-section (2) provides for the imposition of a speed limit on roads equipped with systems of street lighting after 1st July 1957.
(8) Road Traffic Regulation Act 1967, section 72(3).
(9) Motor Vehicles (Variation of Speed Limits) Regulations 1984.
(10) Car-derived vans are goods vehicles derived from a passenger vehicle with a maximum unladen weight not exceeding 2 tonnes.

Class of Vehicle	Maximum Permitted Speed when used on		
	Dual Carriageways	Other Roads	Motorways
Drawbar trailer outfit over 7500 kg GPW	50 mph	40 mph	60 mph*
Motor Tractors and Locomotives (a) complying with requirements as to springs and wings in C & U Regs, if drawing trailers which also comply	30 mph	30 mph	40 mph*
(b) in any other case	20 mph	20 mph	20 mph*
Works trucks	18 mph	18 mph	18 mph*
Vehicles with some or all axles having resilient (non-pneumatic) tyres	20 mph	20 mph	20 mph*
Vehicles having any wheel with non-resilient non-pneumatic tyres	5 mph	5 mph	5 mph*
Vehicles constructed or adapted for the carriage of Abnormal Indivisible Loads (11)	12 mph	12 mph	60 mph (12)
Vehicles as above which are (a) unladen, (b) able to comply with C & U Regs relating to springs, brakes, tyres, wings and transmitted weight	20 mph	20 mph	60 mph (12)
Engineering Plant	12 mph	12 mph	60 mph (12)(13)
Vehicles authorised (11) to carry loads more than 4.3 metres (14ft) wide but less than 6.1 metres (20ft) wide	20 mph	20 mph	60 mph (12)

* - Vehicles over 7.5 tonnes GVW or over 3 tons unladen weight, and vehicles drawing trailers, are banned from the third (outside) lanes on Motorways.

<u>Speeding</u> is an offence carrying obligatory endorsement of the driver's ordinary driving licence. In addition, as well as the endorsements counting towards disqualification under the 'totting up' rules (3 penalty points), courts have discretionary powers of disqualification in connection with the offence (14).

A timetable or schedule issued by an employer requiring a journey to be completed in less time than is practical legally to do so, may be used as evidence that the

---

(11) Motor Vehicles (Authorisation of Special Types) Order 1973 (Amendment Order made 1971).
(12) 70 m.p.h. in the unlikely event that they are under 3 tons unladen weight.
(13) Not permitted on Motorways unless they are able to attain at least 25 m.p.h.
(14) Road Traffic Act 1972, section 22 and schedule 4.

employer procured or incited the employee to commit a speeding offence (15).

5.1.2 Parking, Waiting, Loading and Unloading

The everyday use of a goods vehicle on a road, particularly when it is stationary for loading and unloading purposes, can create situations which may cause the user to offend against various statutes concerned with parking, waiting, obstruction, and leaving a vehicle unattended. The chief of these regulations are to be found in the Road Traffic Acts (1), the Road Traffic Regulation Act (16) and the Construction and Use Regulations.

In addition, many local Acts and Orders permit vehicles to stand in 'No Waiting' streets for specific purposes, e.g. loading or unloading, or to permit a person to board or alight. In fact is more usual to restrict loading and unloading to certain times, than to ban it completely. In the event of any doubt arising, the relevant local Order must be scrutinised; but in the vast majority of cases the information is given by an adjacent sign of a prescribed type (17).

No Waiting This restriction prohibits the parking of vehicles. The existence of such a prohibition is indicated by the usual 'No Waiting' sign of a red circle, infilled with blue, with a red diagonal. The extent of the ban is indicated:-
(i) by yellow markings along the edge of the carriageway -
- (a) a broken line indicates a peak hour ban;
- (b) a single solid line indicates a working day ban, and this may include Sundays;
- (c) a double solid line indicates a ban for longer than the working day;

(ii) by adjacent signs giving the days and times during which waiting is prohibited or limited.

However, as explained above, this restriction is usually tempered by a local order permitting loading and unloading. The extent to which these local orders can vary is considerable.

If a driver parks while he goes to ascertain at adjacent premises if there is a load for him, he may offend against a local order which permits waiting only while loading (18). However, if the driver leaves his van while he delivers goods, he may be said to be engaged in unloading and therefore within the scope of the exemption (19).

To stop to pick up an easily portable object (20), e.g. a parcel, is not loading. Where stopping to allow boarding and alighting is permitted by a local order (21), it goes beyond the terms of this exemption to stop to deliver a parcel to nearby premises. The exemption for loading and unloading, where it occurs, does not extend to a driver stopping for his statutory break.

No Loading or Unloading - No Parking If the above 'no waiting' restriction is further qualified by a ban on loading and unloading, it effectively constitutes an absolute prohibition against parking.

The existence of such a ban is shown by yellow markings on the kerb at right

---

(15) Road Traffic Regulation Act 1967, section 78A; Road Traffic Act 1972, section 203.
(16) The Road Traffic Regulation Act 1967.
(17) Traffic Signs Regulations and General Directions 1964.
(18) Holder v. Walker 1964.
(19) McLeod v. Wajkowska 1963.
(20) Sprake v. Tester 1955.
(21) Clifford Turner v. Waterman 1961.

angles to the carriageway; the extent of the ban is indicated by adjacent signs giving the days and times during which loading and unloading is prohibited:-

(a) A single yellow line repeated at intervals indicates a peak hour ban;
(b) A double yellow line repeated at intervals indicates a working day ban, which may include Sundays;
(c) Three yellow lines repeated at intervals indicates a ban for longer than the working day.

The Police are authorised to permit loading and unloading outside the time limits shown. Prior or immediate permission can often be obtained where circumstances dictate stopping at a banned kerb.

Clearways and Motorways There is a general prohibition against stopping on Clearways and Motorways, subject to the following:-

(a) Permission may be obtained from a Police Constable in uniform to stop on a Clearway (22);
(b) A vehicle may be driven on to the 'hard shoulder' of a Motorway only for as long as necessary in the following circumstances:-
  (i) Breakdown, including lack of fuel, oil or water;
  (ii) Accident;
  (iii) Illness or other emergency;
  (iv) To render assistance to another person legitimately stopped (23).

Stationary Vehicles - Driver's Duties Three of the Construction and Use Regulations concern stationary vehicles.

Regulation 117 imposes an obligation on the driver of a vehicle which is stationary (other than for an enforced traffic stop) to stop the engine, unless this is required to work some ancillary equipment, e.g. a tanker's delivery pump, through a power take off. The following Regulation 118 prohibits the driver from sounding his horn while the vehicle is stationary.

Under Regulation 124, if the driver leaves the vehicle unattended he must set the handbrake and switch off the engine.

Obstruction To unreasonably deprive other road users of road space, in such a manner as to hamper their passage, constitutes obstruction. There would appear to be degrees of obstruction, e.g. wilful obstruction (24) and unnecessary obstruction (25); in fact merely to leave a vehicle parked for an unreasonable length of time may amount to obstruction (26). It is not necessary for the prosecution to prove that actual obstruction occurred; potential obstruction can be shown (27). A vehicle may obstruct a footpath (28), but in an emergency, or where prior permission has been obtained from a Police Constable in uniform, or where for the purposes of loading and unloading it would have been unsatisfactory to park elsewhere and the vehicle is not left unattended, it is permissible to park on a verge, footway (29) or central reservation (30).

A similar offence to obstruction, that of **leaving a vehicle in a dangerous**

---

(22) Various Trunk Roads (Prohibition of Waiting) (Clearways) Regulations 1962. On urban clearways, adjacent signs indicate the extent of loading bans.
(23) Motorways Traffic Regulations 1959, section 7.
(24) Highways Act 1980, section 137.
(25) Construction and Use Regulations 1978, section 122.
(26) Soloman v. Durbridge 1956.
(27) Gill v. Carson & Nield 1917.
(28) Bryant v. Marx 1932.
(29) Road Traffic Act 1974, section 7. The Act amends section 36 of the Road Traffic Act 1972. See also (30) following.
(30) Heavy Commercial Vehicles (Control and Regulation) Act 1973.

position, i.e. so that it is likely to cause danger to other road users, is contained in the Road Traffic Act (31).

If a vehicle is parked at night, other than with its nearside as close as possible to the edge of the carriageway, the user may offend not only against the above Act, but also against the Construction and Use Regulations (32).

Parking at Night without Lights The circumstances under which it is permissible to park without lights during the hours of darkness are fully set out in Chapter 8 - section 8.4 (Lighting and Marking).

Pelican and Zebra Crossings Stopping is prohibited on the approaches to Pedestrian Crossings, unless the vehicle is halted at the stop line to allow pedestrians to cross (33). However, it is a good defence to show that the driver stopped because of circumstances beyond his control, or to avoid an accident. Overtaking is also prohibited within the 'zebra control zone' (34).

5.1.3 Traffic Offences

Permitting, Causing and Using The driver is not always the only person who may be prosecuted for a road traffic offence; the law may also proceed against the person who, by his express or implied permission, **permits** (35) the driver, albeit at his discretion, to use the vehicle in contravention of the Acts and Regulations, or who **causes** (36) such use by an instruction, mandate or authority. The **user** (37) of the vehicle is the driver or the person whose servant or agent the driver is; but there is a defence available to an employee driver who uses a vehicle in ignorance of the fact that his employer has not insured it (38).

Control of Heavy Commercial Vehicles - 'Lorry Routes' Local Authorities may make traffic regulation (39) orders specifying through routes for heavy commercial vehicles (40), or to prohibit or restrict such vehicles from particular zones or roads. Failure to comply is a ground for revocation, curtailment, suspension or premature termination of an Operator's Licence.

Owner's Liability The registered owner of a vehicle is liable for certain minor traffic offences (41) which are the subject of fixed penalty system ticket fines

---

(31) Road Traffic Act 1972, section 24.
(32) Construction and Use Regulations 1978, section 123.
(33) Pelican Pedestrian Crossing Regulations and General Directions 1969; 'Zebra' Pedestrian Crossing Regulations 1971.
(34) Marked by broken zig-zag lines at the approach to the crossing.
(35) James v. Smee 1954.
(36) McLeod v. Buchanan 1940. Note the defence of 'due diligence' which is available to an employer charged with failing to cause records to be kept. Transport Act 1968, section 98(4).
(37) Transport Act 1968, section 92(2); Ready Mixed Concrete (East Midlands) Ltd v. Yorkshire Traffic Area Licensing Authority, 1970; and Elliot v. Gray 1959.
(38) Carpenter v. Campbell 1953.
(39) The Heavy Commercial Vehicles (Control and Regulation) Act 1973 (The 'Dykes Act').
(40) A Heavy Goods Vehicle is a vehicle constructed or adapted for the carriage of goods, and over 7.5 tonnes Gross Plated Weight (Road Traffic (Drivers' Ages and Hours) Act 1975). Such a vehicle over 3 tons unladen weight is also a Heavy Motor Car (Road Traffic Act 1972, section 190) and thus a Heavy Commercial Vehicle.
(41) Road Traffic Act 1974, sections 1-5.

(42). Hire firms can avoid this responsibility by requiring the hirer to sign an acknowledgement of liability (43). There are procedures whereby the registered owner may discharge his liability. On receipt of a fixed penalty ticket, if he does not pay the fine within 21 days (44), he will receive within 6 months a statement setting out the details of the alleged offence, and obliging him to either pay the fine, or to complete a statutory statement of facts giving the identity of the responsible owner.

Traffic Wardens in uniform aid Police in carrying out certain traffic regulation functions and in traffic law enforcement. Failure to comply with their directions when they are so acting is an offence (45). They may make enquiries to find the identity of drivers of vehicles in certain cases (46), operate the fixed penalty system, and control traffic, in addition to their more usually understood function of street parking and parking meter attendants.

Stealing and Taking Away Vehicles A new offence of vehicle interference with intent to either (a) take and drive away the vehicle, (b) steal the vehicle or (c) steal the contents, is established by the Criminal Attempts Act 1981. The penalty is three months' imprisonment, or a £500 fine, or both.

5.1.4 Safe Loading

The load carried by a vehicle must at all times be so secured and distributed that it does not constitute a danger (47). Responsibility for the safety of the load belongs to the driver, even if the vehicle is loaded for him (48). A Code of Safe Loading Practice (49) applicable to all loads, and the Chemsafe Code (50) for hazardous loads, are published. Whilst the provisions of these codes are not mandatory, contravention of them if proven would obviously be taken into account (51) in any civil or criminal proceedings arising out of the use of an unsafely loaded vehicle.

---

(42) Road Traffic Regulation Act 1967, section 80; Transport Act 1968, section 131(1).
(43) Road Traffic Act 1974, section 3.
(44) 7 days in the case of a parking meter excess charge (Road Traffic Regulation Act 1967; Transport Act 1968, section 127).
(45) Road Traffic Regulation Act 1967, as amended by Transport Act 1968.
(46) The owner of the vehicle has a duty if so requested on behalf of a Chief Officer of Police to give information as to the driver of a vehicle (Keeper Liability; Road Traffic Act 1972, section 168).
(47) Construction and Use Regulations 1978, section 97.
(48) National Coal Board v. Gamble 1959.
(49) Published by HMSO, 30p.
(50) Includes details of the TREMCARD and HAZCHEM systems of labelling and describing loads in transit.
(51) Health and Safety at Work etc. Act 1974.

6 Accidents and Insurance

The Road Traffic Acts (1) (2) place certain obligations on the drivers of vehicles involved in accidents. The Acts deal with accidents to persons, animals and vehicles, and with the production of Certificates of Insurance. The driver has different duties and responsibilities under each section.

6.1 Accidents to Persons, Animals, Vehicles and Property

Where, owing to the presence of a motor vehicle or trailer on a road, an accident occurs involving personal injury, injury to an animal, damage to another vehicle (1) or to any other property on or adjacent to the road (2), the driver must:-

(a) stop;
(b) give certain particulars to any person reasonably requiring these;
(c) if for any reason he does not give these particulars to any such person, he must report the accident to the police as soon as possible, and in any case within 24 hours.

In this section the expression 'animal' means any horse, cattle, ass, mule, sheep, pig, goat or dog.

The Act only applies where there is damage to another vehicle or property, or personal injury to another person. The driver is under none of the above obligations if only his vehicle is damaged and/or only he is injured. Injury has been held to include shock (3). The driver must be aware both of the occurrence of the accident (4) and of its consequences, before he can be shown to have committed an offence.

6.1.1 Duty to Stop

This duty involves not merely stopping, but remaining at the scene of the accident long enough to see if there is any person involved who might require the driver's name and address. Thus the duty is still there even if the accident occurred as a result of the stationary presence of the driver's vehicle. Even if the driver reports the accident later, he may still be guilty of failing to stop.

6.1.2 Duty to give certain particulars

The Act lays down the minimum particulars (5) to be given to any person reasonably requiring them (e.g. a Policeman or another party involved in the accident) as being:-

---

(1) Road Traffic Act 1972, section 25 and Part VI.
(2) Road Traffic Act 1974, schedule 6(12).
(3) Bourhill v. Young, 1943.
(4) Harding v. Price, 1948.
(5) Road Traffic Act 1972, section 25(1).

(a) Name and address of the driver;
(b) Name and address of the owner of the vehicle;
(c) Registration number of the vehicle.

If for any reason, for example because there was no-one at the scene of the accident, the driver does not discharge this duty, then he has a:-

6.1.3 Duty to Report the Accident

The accident must be reported as soon as is reasonably possible; if a driver could have reported immediately but waits a few hours, he may still offend even if he reports within 24 hours (6). He may also offend if he refuses his name and address to any person reasonably requiring them, even if he reports the accident to the police within 24 hours (7). If he is interviewed by the police within 24 hours as a result of the accident, this does not constitute 'reporting'.

6.2 Compulsory Insurance

The operator is under legal obligations to insure for third party liability, passenger liability and employer liability.

6.2.1 Third Party and Passenger Liability

It is an offence to use, or to permit the use of, a motor vehicle on a road unless its use is covered by an insurance policy (or security, see below) complying with the requirements of the Road Traffic Act (8) concerning liability in negligence to a third party.

The 'user' of a vehicle is the driver or the person whose servant or agent the driver is (9). However, there is a specific defence if an employee can show that the vehicle did not belong to him, nor was it on hire or loan to him, and that he was using it in the course of his employment with no reason to suspect failure to insure (10). An employee in such circumstances would not be convicted, but this would not prevent the employer being prosecuted for causing or permitting the use of an uninsured vehicle.

Securities An undertaking can, if it wishes, 'carry its own insurance', i.e. accept liability for any claims against it by a third party in respect of the use of its vehicles on a road, if it deposits, and keeps deposited, with the Accountant General of the Supreme Court a sum of £15,000 (11).

To meet the requirements of the Act, the policy must insure against claims by third parties in respect of death or bodily injury <u>to any person</u> arising out of the use of the insured vehicle on a road. It must also cover payment of a medical fee for emergency treatment to a third party by a doctor or at a hospital (12).

Passenger Liability The words **'to any person'** include passengers on the insured

---

(6) He does not, however, have a duty to use **every** means at his disposal, e.g. if there happened to be a nearby phonebox which he ignored - Britton v. Loveday, 1980.
(7) Dawson v. Winter, 1932.
(8) Road Traffic Act 1972, part VI, section 143.
(9) A passenger discovering, after accepting a lift, that the driver is driving without the owner's consent, is not himself 'using' the vehicle - Boldizer v. Knight, 1980.
(10) Road Traffic Act 1972, section 143(2).
(11) Road Traffic Act 1972, Part VI, section 144; Motor Vehicles (Third Party Risks Deposits) Regulations 1967.
(12) Road Traffic Act 1972, sections 145(a)(b) and 154.

vehicle. The exclusion clause for passengers in earlier Road Traffic Acts (13) is omitted in the current Act.

6.2.2 Employer's Liability

An employer has a common law responsibility for the safety of his employees, as well as in many cases for the consequences of their actions, where they are under a contract of service to him. This is known as **vicarious liability**, which is a stricter liability than the duty merely to check that a contract for service is satisfactorily completed.

But the employer also has a statutory duty to insure against claims which arise out of or in the course of employment. He must additionally display the appropriate certificate of insurance at the workplace (14).

The occupier of premises, not necessarily the owner, is under an obligation to all lawful visitors - welcome or unwelcome! - unless they are actually trespassing, to ensure their safety whilst on his premises. Occupier's Liability Cover is available from most insurance companies (15).

6.2.3 Authorised Insurers

A Road Traffic Act policy must be issued by an authorised insurer (16). The main conditions of authorisation are:-
(a) The deposit of £15,000 at the Supreme Court as a security;
(b) Compliance with Board of Trade rules (17);
(c) Membership of the Motor Insurance Bureau (18).

6.2.4 Certificate of Insurance

Requirements of the policy The Policy must be in the prescribed form (19) (20). If the insurance is cancelled, the certificate of insurance must be surrendered to the insurer within seven days (20).

Issue of Insurance Certificates An insurance policy is ineffective until a certificate of insurance is delivered to the policy holder (20) in the prescribed form, but the term 'Policy of Insurance' includes a Cover Note (21). The policy cannot be transferred to a new owner of a vehicle without the insurer's consent

---

(13) Road Traffic Act 1960 section 203(a). The omission of this Section from the Road Traffic Act 1972 section 145, in effect consolidates the Motor Vehicle (Passenger Liability) Act 1971 into the 1972 Act.
(14) Employer's Liability (Compulsory Insurance) Act 1969. The minimum cover is £2 millions. Because of the provisions of this Act, the exemption clause for employees remains in the Road Traffic Acts (Road Traffic Act 1960, section 203(b) now replaced by Road Traffic Act 1972 section 145(a)).
(15) Occupiers' Liability Act 1957. The Health and Safety at Work etc. Act 1974 extends this liability to members of the public affected in any way by the operator's work. Also the Motor Vehicles (Compulsory Insurance) (No.2) Regulation 1973 gives effect to an EEC directive of 1972 requiring vehicle insurance to extend compulsory cover for civil liability, by amending section 145(3) of the Road Traffic Act 1972.
(16) Road Traffic Act 1972, section 152(2).
(17) Companies (Insurers and Partnerships) Act 1967.
(18) Road Traffic Act 1974, section 20.
(19) Motor Vehicles (Third Party Risks) Regulations 1972.
(20) Road Traffic Act 1972, section 147.
(21) Road Traffic Act 1972, section 158(1); Peters v. General Accident Co., 1938.

(22).

Protection of the Third Party's Interests If a third party has a claim against an insured person, and the latter becomes bankrupt, the right of the bankrupt under a Third Party Policy of Insurance (i.e. any settlement of the claim) is vested in the injured third party (22).

An insurer cannot evade his liability to a third party on technical grounds (23), i.e. he cannot negative his liability with what are commonly known as exclusion clauses, such as that the vehicle should be maintained in an efficient condition, because such clauses are rendered void. (But see below under 'Exclusion Clauses').

A third party who is injured as a result of the use on the road of an uninsured vehicle is further protected through the workings of the **Motor Insurance Bureau.** This is an incorporated body of insurers who have agreed to satisfy any claims arising out of such a situation. It should be noted, however, that an uninsured user of a motor vehicle cannot escape liability for using a motor vehicle uninsured, by showing that any claim against him would in any case have been satisfied by the Motor Insurance Bureau (24).

6.2.5 Exclusion Clauses

The provision (above) which prevents any attempt by an insurer to avoid third party liability through exclusion clauses, nevertheless does not prevent an insurer from issuing a policy subject to stated conditions. In other words, the statute lists certain exclusion clauses which will be void in a third party claim and cannot then be relied upon by the insurer in such cases. If an insurer imposes conditions, e.g. as to use, he may still rely on these to exclude his liability to the insured, even though he cannot rely on them to exclude any liability to a third party. Further, since it is by no means clear whether the above provision (23) would be a successful defence, the operation of the exclusion clause could leave a driver open to prosecution for driving without insurance. Thus a motor vehicle insured for domestic and pleasure purposes only will not be insured when used for the carriage of goods for hire or reward, because under the terms of the policy this will **invalidate** the insurance.

In another case, insurers were able to avoid liability to the insured after an accident shown to have been caused by defective brakes (25). Excessive alcohol consumption has been held to breach a contract of insurance (26). Even a defect which no amount of reasonable maintenance could have avoided, was held by a court in 1936 to have breached the condition that the insured vehicle should not be driven in an unroadworthy condition (27); more recently the overloading of a (passenger) vehicle has similarly been held to breach a condition of the insurance (28). Of course, the 1972 provisions (23) above, lessen **but do not remove** the severity of these exclusions.

A growing practice among insurance companies is to issue 'blanket' cover notes to an insured person for the use, subject to the conditions of the policy, of any vehicle of a particular description. Such a policy will be admitted as proof of

---

(22) Third Party (Rights against Insurers) Act 1930; Road Traffic Act 1972, section 150.
(23) Road Traffic Act 1972, section 148.
(24) Monk v. Warbey, 1935.
(25) Liverpool Corporation v. Roberts and Marsh, 1964.
(26) NFU Mutual Insurance v. Dawson, 1941.
(27) Trickett v. Queensland Insurance, 1936.
(28) Clarke v. National Insurance Association, 1964.

insurance when application is made for an excise licence for the vehicle (29).

Danger of Inadvertent Invalidation of the Insurance Policy Two common exclusion clauses in Motor Vehicle Policies which enable an insurer to negative his liability, concern **making a statement at the time of an accident** and **failure to report the accident to the insurer.** Because of a principle of insurance law known as subrogation (i.e. the vesting of the rights of the insured with the insurer), the insurance company, having indemnified the insured, may wish to act as the insured's agent in proceeding against the other party. If the insured or his servant, the driver, has either made a statement admitting liability or failed to report the accident to the insurer, he must place the insurance company at a disadvantage, and to this extent such exclusion clauses can be seen to be reasonable. But their effect, if they are ignored at the time of the accident, could be to invalidate the policy, leaving the driver uninsured.

6.2.6 Production of Certificate of Insurance

The Policy must be produced if required to the Police (30), but if it is not in the immediate possession of the user it may be produced by him (or by his agent (31)), at any Police Station named by him, within five days.

This procedure is similar to the Section 25 provision relating to procedure at the time of an accident, but it imposes only one obligation, i.e. the production of a certificate of insurance, and applies in only one instance, i.e. where there is personal injury to another person. Another person may be a passenger, but does not include the driver. As with Section 25, the obligation is not there if the driver is genuinely unaware of the accident or the injury (4).

The certificate of insurance may be produced to any person having reasonable grounds for requiring it. If this is not done, as soon as is reasonably possible but in any case within 24 hours the accident must be reported to the police and the certificate of insurance produced.

Note also the difference from Section 25, in that it is an offence to fail to produce the certificate of insurance to a policeman, but not an offence to fail to produce it to another person. The driver may elect to have the certificate of insurance produced (by himself or his agent) at such police station as he may specify at the time of reporting, within five days of the accident (31). (Compare this provision with the requirement for **personal** production of a driving licence).

6.2.7 Comprehensive Policies

It should be realised that an operator's legal responsibility under the Act is merely to provide insurance against his liability to third parties arising out of the use of his vehicle. However, he may incur further liability in using his vehicles - for example, he may damage his own or another vehicle, or property such as a shop front. A fully comprehensive policy would cover him against such risks, but even this may include such clauses as a third party damage limit. Additional cover may be desired by the operator, and sought through his broker or insurer.

An operator who carries goods for hire or reward incurs strict liability for the safety of these goods. He may obtain a slight measure of protection from the Carriers' Act of 1830 which applies to 'common carriers'; but as the great

---

(29) Road Traffic Act 1972, section 153.
(30) Road Traffic Act 1972, section 166.
(31) Road Traffic Act 1962, Schedule III.

majority of hauliers carry goods subject to their own conditions of carriage or to those of their trade association (e.g. the RHA), this relieves them of some of the most onerous responsibilities of a common carrier, and makes them in law 'private carriers', able to rely on the above conditions of carriage. However, a carrier whose conditions of carriage disclaimed all liability would soon find himself with nothing to carry. Some form of **Goods in Transit** insurance becomes essential, to cover the carrier for claims against him under his conditions of carriage. This is more than ever true today, when a carrier may, perhaps unwittingly, be party to the carriage of a consignment under the CMR Regulations (32) of 1965, and may incur a proportion of the liability for some damage to the consignment occurring, say, on the Continent. An extension of the Goods in Transit policy to cover this risk is well worth considering.

Some Goods in Transit policies contain **Immobiliser** and **Night Risk** clauses requiring goods vehicles used for high value loads to be fitted with an approved anti-theft device, and to be securely parked and not left unattended unless immobilised or parked in locked premises.

Whilst a 'comprehensive' policy of vehicle insurance will usually cover fire damage and actual theft, it is prudent to seek cover for **fire and theft** losses incurred in the business other than by operation of the vehicles, e.g. at depots.

Staff may be responsible for assets, stock, vehicles and goods in transit (both belonging to the operator and to his customers, e.g. trailers) worth many thousands of pounds. The dishonest employee can cost the operator dearly. For a reasonable premium most insurers will arrange **fidelity bonds** on employees to cover against such an unfortunate eventuality, and it can be well worthwhile for an operator to discuss this type of insurance with his broker. However, it cannot be emphasised too strongly that the operator must accept some responsibility for selecting reliable staff - failure to do so will be reflected all too soon in increased premiums.

---

(32) Carriage of Goods by Road Act 1965.

7 Weights and Dimensions

The major legislation affecting the permitted weights and sizes of vehicles is the Motor Vehicles (Construction and Use) Regulations 1978 which have been made by the Minister of Transport under 'enabling' powers given to him by section 40 of the Road Traffic Act, 1972.

Certain loads which exceed these weights and dimensions may still be carried by virtue of the Motor Vehicles (Authorisation of Special Types) General Order 1979 (1). Many of the provisions of both sets of regulations are either the same or similar, and it is convenient for this reason to study them together.

7.1.1 Definitions

The Regulations refer to Goods Vehicles, and this class of vehicle is further sub-classified for various purposes, e.g. O-licensing, Driver Licensing, and Drivers' Hours and Records regulations. In the main the basis of the sub-classification is increasingly becoming the vehicle's Gross Plated Weight (2). To avoid confusion between what are often very similar terms, the following definitions are provided.

Goods Vehicle - A vehicle constructed or adapted for the carriage of goods, including trailers.

Motor Vehicle - A mechanically propelled vehicle intended or adapted for use on the roads (3).

Trailer - A vehicle drawn by a motor vehicle (4).

Articulated Vehicle - A motor car or Heavy Motor Car (3) with a trailer so attached that 20% of its load (when uniformly distributed) is borne by the drawing vehicle (5).

Motor Tractor - A motor vehicle with an unladen weight under 7 tons 5 cwt., not

---

(1) Also Road Traffic Act 1972, section 42.
(2) Prior to the implementation of the Plating and Testing Regulations, Unladen Weight of the vehicle was the more usual criterion.
(3) Further classified, in the case of a Goods Vehicle, as:-
    (a) Motor Car - below 3 tons unladen weight;
    (b) Heavy Motor Car - above 3 tons unladen weight.
(4) See also under Chapter 9, section 9.3.9 'Composite Trailers'.
(5) An 'artic' is considered as two vehicles for the purposes of operators' licensing and for Class IV (small artic) HGV Driving Licence regulations; and as one vehicle for the purposes of taxation (calculated on the maximum Gross Combination Weight), gross weights and speed limits.

Fig. 2 MAXIMUM TRANSMITTED WEIGHT

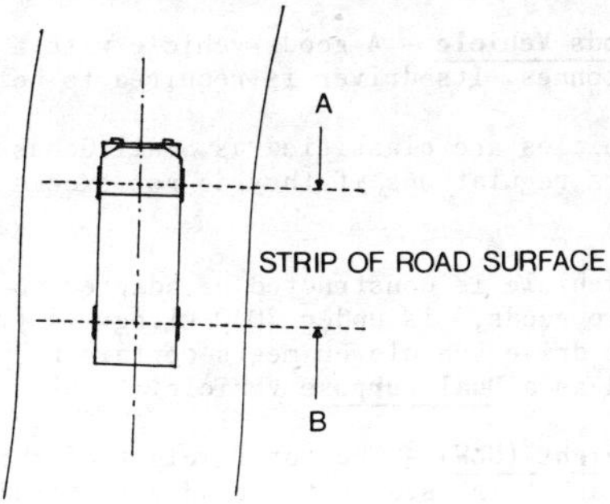

DISTANCE A – B	MAX. TRANSMITTED WEIGHT	REFERENCE
Under 1.02 m 1.02 m – 1.22 m 1.22 m – 2.13 m	11 Tons (see Fig. 3.2) 16 Tons 18 Tons	Construction and Use Regulations 1978, section 94
Under 610mm Each extra metre until 2.13 m Each extra metre until 8.00 m	45720 kg 30000 kg 91320 kg 10000 kg 150000 kg	Authorisation of Special Types Order 1973
Maximum per wheel including 'twins' (see Fig. 3.1b) with centres closer than 610mm apart	11430 kg	

constructed to carry a load (6).

Light Locomotive - A motor vehicle with an unladen weight over 7 tons 5 cwt and under 11 tons 10 cwt, not constructed to carry a load (6).

Heavy Locomotive - A motor vehicle with an unladen weight over 11 tons 10 cwt, not constructed to carry a load (6).

Heavy Goods Vehicle - A goods vehicle with a Gross Plated Weight (see below) of over 7.5 tonnes. Its driver is required to hold an HGV Driving Licence.

Goods vehicles are classified as Small Goods Vehicles for O-licensing and Drivers' Hours regulations if they do not exceed 3.5 tonnes Gross Plated Weight (see below).

If the vehicle is constructed or adapted to carry passengers instead of or in addition to goods, is under 2040 kilograms (2 tons) unladen weight and is either a 4-wheel drive vehicle or meets certain body construction requirements, it is classified as a **Dual Purpose** vehicle.

Unladen Weight (ULW) - The total weight of the vehicle (7) including the body and all parts as it is used on the road but excluding water, fuel, spare wheel, tools and loose equipment.

Gross Vehicle Weight (GVW) - The total weight of the vehicle and its load which is transmitted by the wheels to the road surface.

Gross Train Weight - Maximum laden weight of a vehicle and trailer.

Gross Combination Weight - Maximum laden weight of an articulated combination.

Gross Plated Weight - Maximum permitted weight at which the vehicle may be used on the road.

7.1.2 Transmitted Weights and Axle Weights

Both the Construction and Use Regulations and the Special Types Order lay down maximum weights which can be transmitted to any strip of road surface at right angles to the longitudinal axis of a vehicle (Fig. 2). The much higher maxima are only allowed under the Special Types Order subject to an indemnity (8) being given. No Abnormal Indivisible Load or Engineering Plant may cause a Special Type Vehicle to exceed 150 tons (152,400 kg) GVW. No single wheel must carry more than 11 tons (11,430 kg).

The transmitted weight regulations, together with the C & U Regulations, limit the weight transmitted by a single wheel where no other wheel is in line transversely to 5 tons, and limit the weight transmitted by two wheels in line transversely to 10 tons. A sole driving axle on an articulated combination exceeding 32 tons (32,520 kg) may transmit a weight of 10,500 kg. The weight limit on a steerable wheel is 3.5 tons. The effects of these regulations can be seen in Fig. 3 (Maximum permitted axle loadings).

In this context also it should be noted that the Plating Regulations may have the effect of prescribing lower maximum axle weights for certain vehicles.

---

(6) i.e. drawing load-carrying vehicles.
(7) For registration purposes, the Unladen Weight is applicable to Goods Vehicles under 1525 kg ULW (Private and Light Goods Vehicles).
(8) Against damage caused to any highway or bridge over which the vehicle passes.

Fig. 3 MAXIMUM PERMITTED AXLE LOADINGS

1. Single Axles

a) Steering

The Maximum Permitted Weight on any steerable wheel is 3.5 tons.
(i.e. 7 tons per axle)

b) In Line

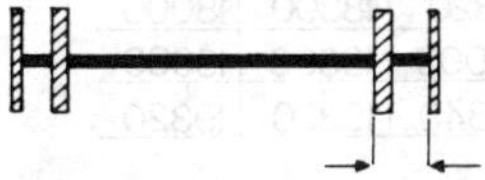

5 tons (5090 kg) per wheel (ie 10 tons 10170 kg max), but in practice since each tyre will be unable to support over 3·5 tons, wide profile (over 300mm) tyres are used, or twin tyres with centres more than 300mm apart (which then count as single wheels). The sole driving axle of an articulated combination exceeding 32 tons (32520 kg) may be loaded to 10500 kg (10·5 tonnes).

2. Oscillating Axles

"4-In-Line"

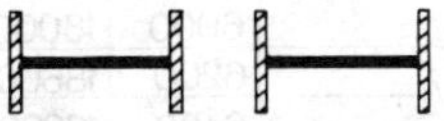

11 tons max [see note (a) Fig. 1]

Fig. 4 SINGLE, TANDEM AND TRI-AXLES

Single Axle.

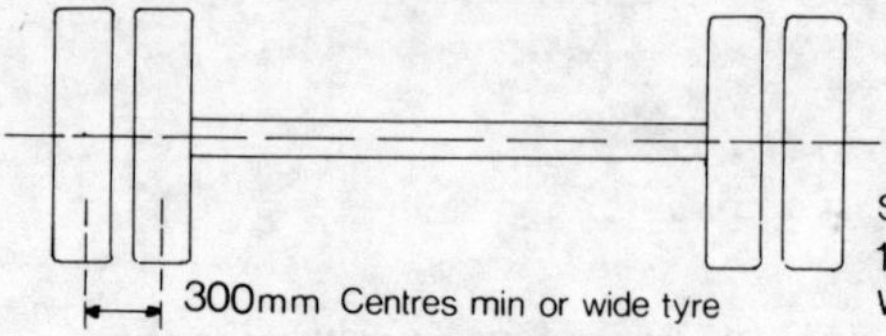

SINGLE AXLE 10170Kg SOLE DRIVING AXLE 10500Kg – PROVIDING a RELEVANT TRAIN WEIGHT OF MOTOR VEHICLE EXCEEDS 32520Kg

Two Closely Spaced Axles.

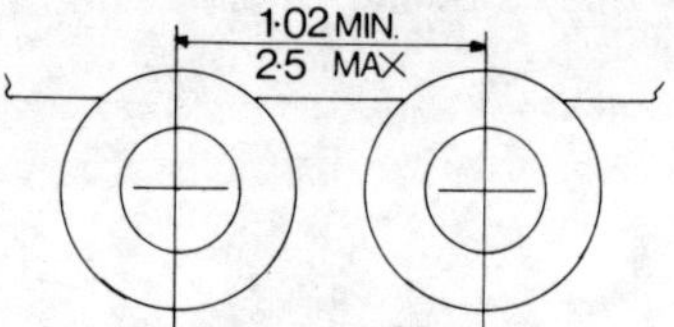

	Max bogie weights Kgs I	II	III
At least 1·02M	16260	12200	10500
At least 1·05M	17280	15260	10500
At least 1·20M	18300	16270	15260
At least 1·35M	18800	17280	16500
At least 1·50M	19320	18300	18000
At least 1·80M	20000	19000	19000
At least 1·85M	20340	19320	19320

I PLATED AXLE WEIGHT EQUAL
II PLATED AXLE WEIGHTS NOT EQUAL, NEITHER AXLE EXCEEDING 10170Kgs
III BOGIE WEIGHT IN CASES NOT WITHIN COLUMN I OR II
eg 3 AXLE MOTOR VEHICLE WITH SOLE 10500Kg DRIVE AXLE

Three Closely Spaced Axles.
No axle plated for more than 7500kg.

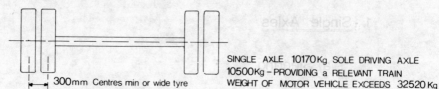

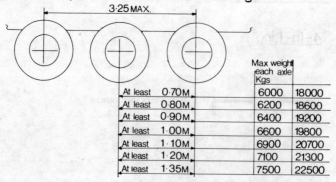

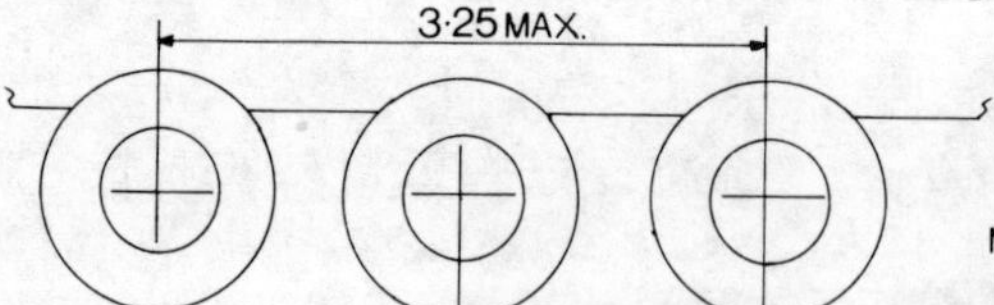

	Max weight each axle Kgs	
At least 0·70M	6000	18000
At least 0·80M	6200	18600
At least 0·90M	6400	19200
At least 1·00M	6600	19800
At least 1·10M	6900	20700
At least 1·20M	7100	21300
At least 1·35M	7500	22500

The weight which can be transmitted by tandem axles and tri-axles is governed by tables of bogie weights contained in the Construction and Use (Amendment No. 7) Regulations 1982 (S.I.1576). The tables are logical in that the greater the axle spread, the higher the permitted bogie weight. The Tri-axle table simply allows higher weights (given in terms of maximum loads on any axle) in proportion to the strip of road surface on which the bogie rides (i.e. the axle spread); but the Tandem axle table distinguishes between bogies with compensating axles (9), bogies with offset axles (10) where neither axle weight exceeds 10 tons (10,170 kg), and bogies outside these limits (for example, a bogie containing a 10,500 kg driving axle), and gives maximum total bogie weights. (See Fig. 4) (11).

Semi-trailers with **three adjacent wide-spaced axles** may have an imposed axle weight up to 24 tons (24,390 kg) in accordance with the following schedule.

Distance between first and third axles	Intermediate (middle) maximum axle weight	Maximum total axle weights (all three axles)
Under 3 metres	not specified	18,290 kg
At least 3 metres	8,390 kg	20,330 kg
At least 3.8 metres	8,640 kg	22,360 kg
At least 4.6 metres	9,150 kg	24,390 kg

7.1.3 Gross Vehicle Weights

The maximum permitted gross vehicle weights of goods vehicles, combinations of goods vehicles and articulated vehicles are governed by schedules of the Construction and Use Regulations (12). These schedules are in fact a 'bridge formula' under which, very approximately, it will be seen that the wider the axle spacings, the greater the permitted GVW. The main provisions of the schedules are shown diagramatically in Fig. 5.

The tables in the Construction and Use Regulations are more complicated than Fig.5, in so far as they allow certain GVWs to be achieved at slightly shorter axle spreads where prescribed limits are observed on the maximum load to be imposed on all axles (or, in some cases, the intermediate axle) (13).

In the case of rigid vehicles, the important measurement is the outer axle spread, and in the case of articulated combinations, it is the relevant axle spread between the rear axle of the drawing unit and the rear axle of the semi-trailer.

The maximum Gross Combination Weight of a 5-axle articulated combination was increased to 38,000 kg in 1982 (14). Figs. 6 and 7 (11) show how these higher weights can be achieved. Note that a new additional criterion, minimum overall length, applies above 32,520 kg. Tractive units can be 2-axle or 3-axle, and must be first used on or after 1st April 1983. The maximum Gross Train Weight of a

---

(9) i.e. with equal plated weights on each axle.
(10) i.e. with unequal plated weights on each axle.
(11) The Author is indebted to York Trailers Ltd of Corby for permission to use their drawing.
(12) Motor Vehicles (Construction and Use) Regulations 1978, sections 85-92 and schedules 6 and 7; as amended by the Construction and Use (Amendment) (No. 7) Regulations 1982.
(13) Motor Vehicles (Construction and Use) Regulations 1978, section 92 and Schedule 7.
(14) As a result of the recommendations of the Armitage Committee of Inquiry into Lorries, People and the Environment, the Construction and Use (Amendment) (No.7) Regulations 1982 implement these changes.

Fig. 5 GROSS VEHICLE WEIGHTS AND AXLE SPACINGS

REF: (Construction & Use Regulations, 1978, Schedules 6/7).

Vehicle	Diagram	AT LEAST	BUT LESS THAN	GVW/GCW kg MAX
FOUR WHEELED RIGID TRUCK	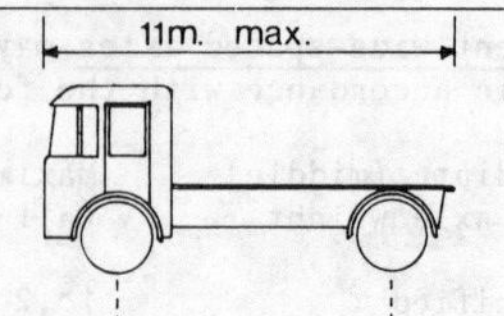		2·65m	14230
		2·65m		16260
		3m		18000 ①
SIX WHEELED RIGID TRUCK	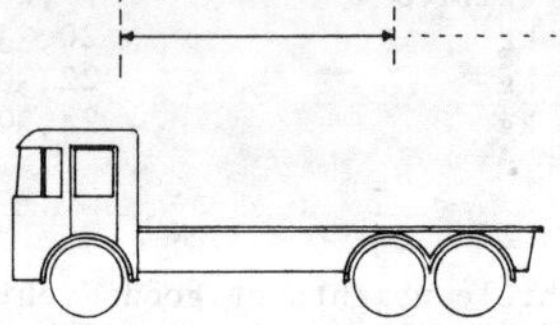		3m	16260
		3m	3·9m	18290
		3·9m	4·6m	20330
		4·6m	5·1m	22360
		5·1m		24390
EIGHT WHEELED RIGID TRUCK	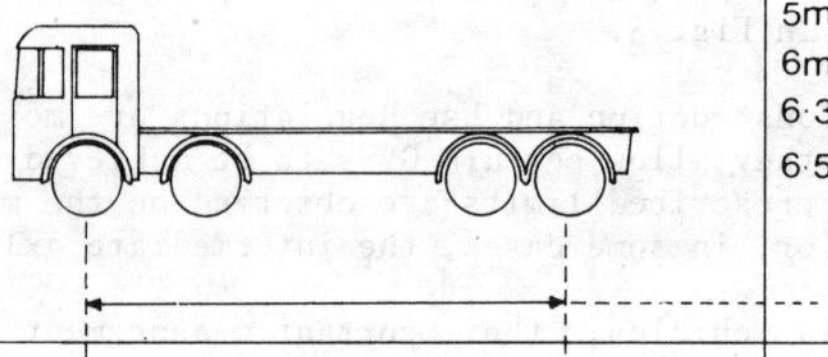	5m	6m	24390
		6m	6·3m	26420
		6·3m	6·5m	28450
		6·5m		30490
FOUR WHEELED TRACTIVE UNIT WITH SINGLE AXLE SEMI-TRAILER	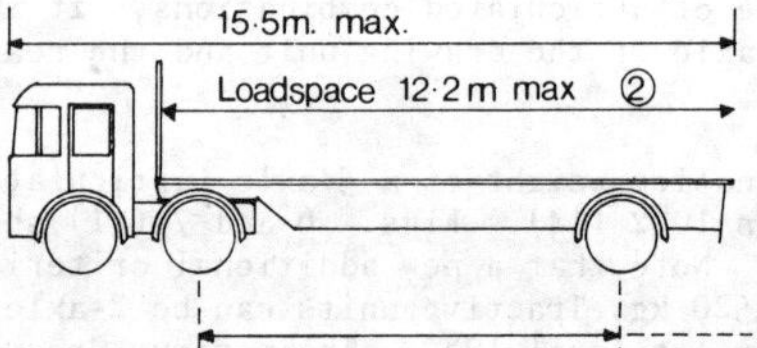	2·0m	2·2m	20330
		2·2m	2·9m	22360
		2·9m		24390
FOUR WHEELED TRACTIVE UNIT WITH TANDEM OR TRI-AXLE SEMI-TRAILER	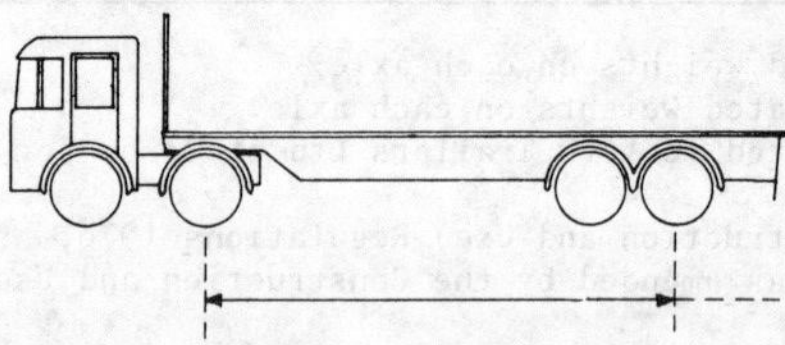	2·9m	3·5m	24390
		3·5m	4·1m	26420
		4·1m	4·7m	28450
		4·7m	5·3m	30490
		5·3m		32520

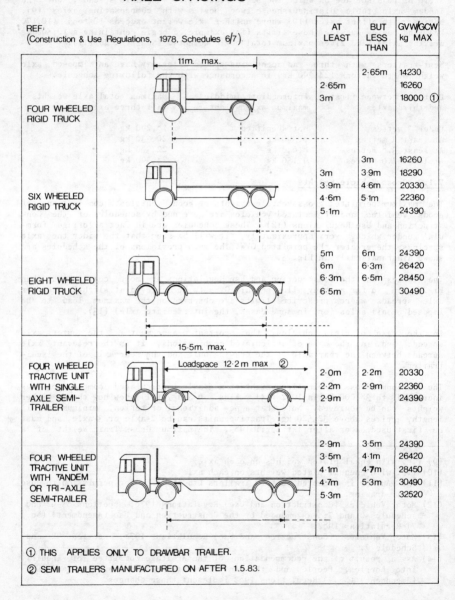

① THIS APPLIES ONLY TO DRAWBAR TRAILER.

② SEMI TRAILERS MANUFACTURED ON AFTER 1.5.83.

Fig.6 ARTICULATED VEHICLES— 5-AXLE COMBINATION, 2-AXLE TRACTIVE UNIT

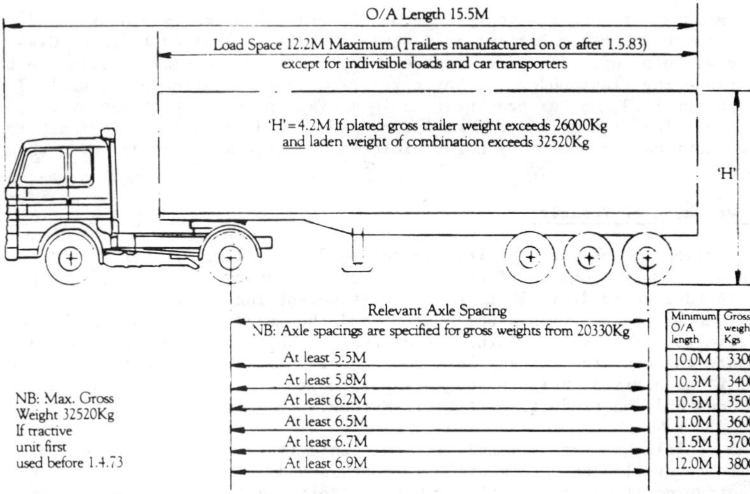

Minimum O/A length	Gross weight Kgs
10.0M	33000
10.3M	34000
10.5M	35000
11.0M	36000
11.5M	37000
12.0M	38000

NB: Max. Gross Weight 32520Kg If tractive unit first used before 1.4.73

Fig.7 ARTICULATED VEHICLES—5-AXLE COMBINATION, 3-AXLE TRACTIVE UNIT

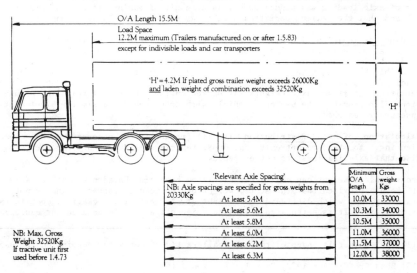

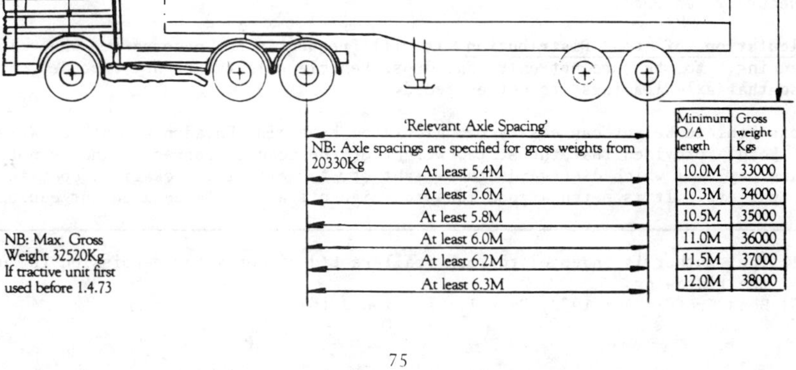

Minimum O/A length	Gross weight Kgs
10.0M	33000
10.3M	34000
10.5M	35000
11.0M	36000
11.5M	37000
12.0M	38000

NB: Max. Gross Weight 32520Kg If tractive unit first used before 1.4.73

draw-bar train is 32,520 kg, but the trailer may have a maximum Gross Plated Weight of 18,000 kg. A 2-axle tractive unit with a relevant train weight over 32,520 kg, and a wheelbase of at least 2.7 metres, may now have a plated weight (for the unit) of 17,000 kg, to facilitate 38 tonne operations with a 3-axled semi-trailer. Separate schedules apply to tractive units (Figs. 8 and 9 (11)).

It is important to realise that the schedule only shows maximum GVWs, and that subject to Plating Regulations many vehicles will be limited to lower GVWs for operational purposes. Also, certain vehicles carrying Abnormal Indivisible Loads and Engineering Plant under the Motor Vehicles (Authorisation of Special Types) General Order 1979, may exceed these maximum GVWs, provided that the owner of the vehicle notifies Highway and Bridge Authorities, and gives a twelve month indemnity against damage caused by his vehicle to any Highway or Bridge over which it passes.

7.1.4 Weighing of Vehicles

Under Section 160 of the Road Traffic Act 1972, a Police Officer who is authorised in writing by a Chief Officer of Police may require a vehicle to stop, and order the driver to drive it to the nearest weighbridge. If the vehicle is more than five miles away from a weighbridge and the weight is found to be within legal limits, the Highway Authority on whose behalf the direction was made must compensate the operator for any loss so caused. A DTp Goods Vehicle Examiner, a specially authorised Weights and Measures Inspector or a specially authorised Police Constable on production of their authority may exercise the same powers.

When the vehicle has been weighed, a certificate is issued; and if the vehicle is found to be overloaded a form GV160 will be issued. This prohibits the vehicle from continuing its journey until it has been offloaded to the required limits.

If the vehicle is found to be overloaded, it is a defence (15) to show that either:-

(a) the driver was proceeding to the nearest weighbridge or, having been weighed, to the nearest practical point at which to offload his surplus weight; or
(b) the overweight occurred after loading, e.g. in falling snow, and in the case of axle loading was caused purely as a result of taking off part of the load, and that the maximum permitted weight was not exceeded by more than 5%.

Loading of Vehicles The Construction and Use Regulations make it an offence to use on a road a vehicle which is unsuitable for the purpose for which it is being used, or is so maintained or loaded as to be a danger to other road users. This is a 'blanket' regulation which might be held to prohibit such things as insecure loading, or unsafely distributing a load on a vehicle so as to exceed the maximum permitted gross and axle weights (as marked on the Ministry Plate). The Regulations also prohibit the use of a vehicle with crane hooks and similar implements insecurely fastened.

Calculation of Load Distribution It will frequently be necessary, when load planning, to be sure not only that Gross Vehicle Weights are not exceeded, but also that axle loadings are not exceeded.

Gross Vehicle Weight can easily be calculated by adding Unladen Weight to Weight of Load, provided that the stated weight of the load is correct, and is not a 'tare payload' which disregards the weight of, for example, packaging, containers or pallets. It is often a good idea to weigh off a vehicle unladen but equipped

---

(11) The Author is indebted to York Trailers Ltd of Corby for permission to use their drawing.
(15) Road Traffic Act 1972, Part II, section 40(6).

Fig.8 MAXIMUM WEIGHT — 2-AXLE TRACTIVE UNIT

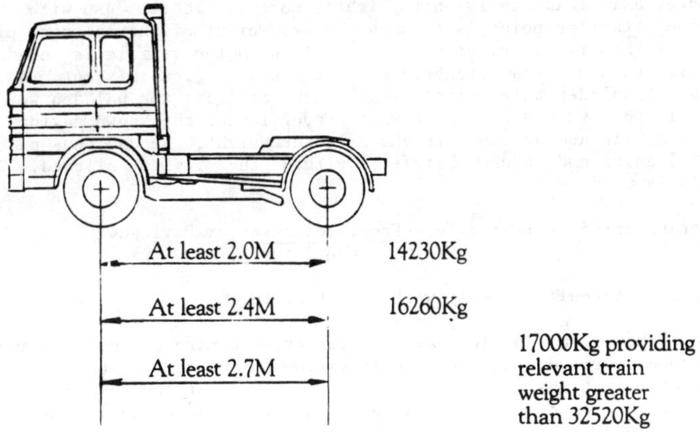

Fig.9 MAXIMUM WEIGHT — 3-AXLE TRACTIVE UNIT

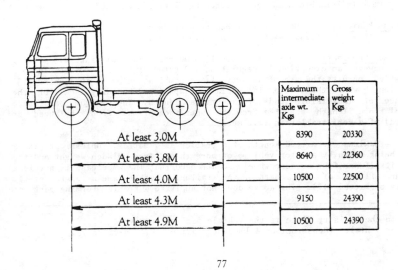

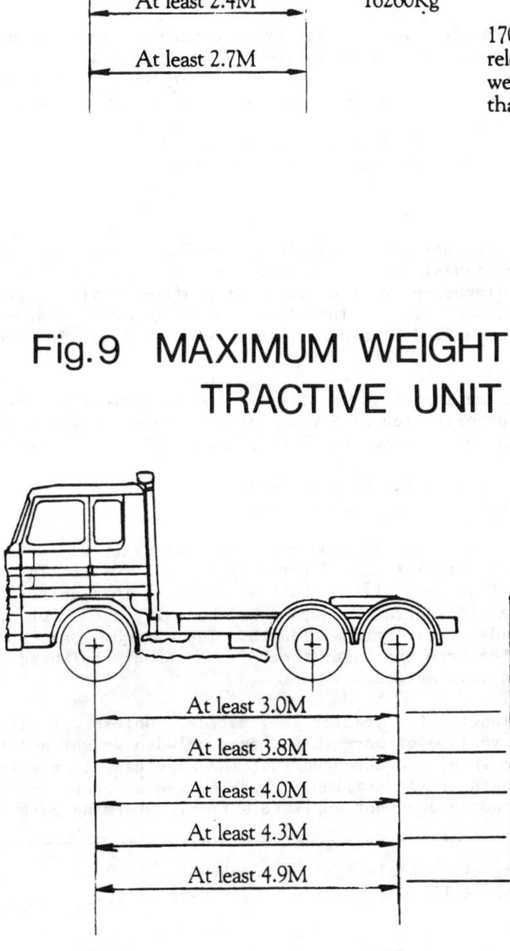

Maximum intermediate axle wt. Kgs	Gross weight Kgs
8390	20330
8640	22360
10500	22500
9150	24390
10500	24390

with spare wheel, tools, sheets and ropes and a full tank of fuel, to obtain a true unladen weight figure. The figure painted on the side of the vehicle (16) is for taxation purposes and does not include these items.

The calculation of axle loadings is only slightly more difficult. As with GVW calculations, the starting point is to know the kerbweight of the chassis plus body, and also how this is distributed on each axle when the vehicle is unladen. Manufacturer's literature or a weighbridge can be used to find this out. As a check, the total of unladen axle weights should give at least the unladen weight of the vehicle. If then we can determine what proportion of the known payload is transferred to each axle and add this to the axle kerbweight, we can thus calculate the axle loading and ensure that it is within the limits allowed. The following formula is used:-

$$\text{Weight of load transferred to rear axle} = \frac{\text{Front loadbase}}{\text{Wheelbase}} \times \text{Payload.}$$

An example using this formula is fully worked in Fig. 10.

It will be obvious that if a vehicle leaves an operating centre correctly loaded, it can still become overloaded on one axle by virtue of taking off some of the load, thus effectively transferring the centre of gravity of the remaining load in the direction of one or other of the axles. This will cause the loadbase to alter and in consequence the proportion of the payload borne by a particular axle will alter, and perhaps exceed the permitted maximum. The defence (b) given above, relating to overloading, was inserted for this very reason.

7.2 Dimensions

The regulations concerning the dimensions of vehicles, combinations of vehicles and articulated vehicles, are amongst the most complex in the C & U/Special Type Regulations. Generally the dimensions of the load must conform to the overall dimensions of the vehicle(s), but wide or long loads which project sideways, forwards or rearwards may be carried within certain limits defined in these regulations.

Many of these limits are the same in both sets of regulations, but some higher limits are allowed under the Authorisation of Special Types Order, particularly where Abnormal Indivisible Loads or Engineering Plant are carried.

Rigid Goods Vehicle, i.e. Motor Car or Heavy Motor Car:-
Maximum length - 11 metres. Maximum width - 2.5 metres.

Articulated Combination - Maximum length 15.5 metres. Maximum width 2.5 metres. There is however an exception in Regulation 140(6)(e) of the C & U Regulations 1978, which permits the use of an articulated combination up to 16.8 metres (55 ft) in length where the trailer is specially adapted and normally used for the conveyance of indivisible loads of exceptional length. The maximum load space (i.e. platform length measured from the inner face of the front bulkhead cr headboard) of a semi-trailer is 12.2 metres.

Trailer - Maximum permitted length of a trailer is 7 metres, unless it is a trailer being drawn by a motor vehicle of more than 2 tons unladen weight and the trailer itself has four or more wheels so arranged that the wheelbase is not less than three-fifths of the length of the trailer; in this case a length of 12 metres is permitted (17). The above does not apply to a trailer forming part of

---

(16) Or on the Ministry Plate if the vehicle is Type Approved.
(17) Motor Vehicles (Construction & Use) Regulations 1978, section 73.

an articulated combination, or being specially constructed and normally used for the conveyance of Abnormal Indivisible Loads.

Maximum permitted width of a trailer is normally 2.3 metres, but it may be 2.5 metres in the case of a trailer being drawn by a motor vehicle of over 2 tons unladen weight, where the drawing vehicle and the trailer both have pneumatic tyres, and the amount by which the trailer, when following straight behind the drawing vehicle, projects laterally at either side, does not exceed 305mm (1ft) (18). The above also applies to a semi-trailer being drawn by a motor car of over 2 tons unladen weight.

Tractor and Locomotive - Maximum permitted length 11 metres. Maximum permitted width - Tractor 2.5 metres, Locomotive 2.75 metres.

Overhang of vehicle body (19) - The maximum permitted overhang of a Goods Vehicle's body is a distance equivalent to 60% of its wheelbase. Overhang is the distance between the rear axle (20) and the back of the body. A container, however, has been held to count as part of the body for the purpose of calculating the overhang (21). An absolute limit of 1.83 metres (6ft) overhang is prescribed for Motor Tractors (which do not, of course, carry any load).

7.2.1 Overall Length (Of the vehicle and its load)

Both the Construction and Use Regulations and the Special Types Order refer to the overall length of vehicles, vehicle combinations and articulated combinations and their loads.

The overall length of a vehicle and its load is counted as the overall length of the vehicle **plus** any forward or rearward projections of the load.

The overall length of a vehicle combination is counted as the sum of:-
(a) The overall lengths of the vehicles **plus**
(b) The distances between them **plus**
(c) The lengths of any projections of the load forward of the foremost vehicle or rearward of the rearmost vehicle.

There is an absolute prohibition on any vehicle combination exceeding 27.4 metres (90ft) in length, and where such a combination exceeds 25.9 metres (85ft) in overall length the operator must:-
(a) Give two clear days' notice to the Police (22); and
(b) Supply an attendant to accompany the vehicle.

Where a load is carried on one vehicle or in such a manner that part of its weight rests on more than one vehicle, whether or not forming part of an articulated combination (23), then if the overall length of the carrying vehicle(s) plus the forward and rearward projections beyond the front of the foremost

---

(18) Motor Vehicles (Construction & Use) Regulations 1978, section 74.
(19) Not applicable to loads, which are said to project rearwards. The distinction between overhang and projection is of fundamental importance in understanding these complex regulations.
(20) In the case of a tandem rear axle, the measurement is from a point 110mm behind the centre point of the tandem axle.
(21) Brindley v. Willett, 1981; Hawkins v. Harold A Russett Ltd, 1983.
(22) Excluding public holidays, Saturdays and Sundays (Motor Vehicles (Authorisation of Special Types) General Order 1979).
(23) Motor Vehicles (Construction & Use) Regulations 1978, section 140(4)(a). This regulation, originally amendments 979/980 to the Construction & Use Regulations 1969, codified the decision in Stevens v. R., 1971.

carrying vehicle and behind the rear of the rearmost carrying vehicle respectively exceeds 18.3 metres (60ft) then the operator must:-
(a) Give two clear days' notice to the Police; and
(b) Supply an attendant to accompany the vehicle.

Doubts about whether the drawing unit of an articulated combination needs to be taken into account when determining the overall length of the vehicle and the load (i.e. the 'sixty-foot rule') are dispelled by sections 139(d)(ii) and 140(4)(a). The former defines the forward projection of a load carried 'in such a manner that part of its weight rests on more than one vehicle' as that part of the load extending beyond the foremost part of the foremost vehicle by which the load is carried, unless the context otherwise requires - and, in the context of the latter Regulation 140(4)(a) (**Overall length of vehicle and load not to exceed 18.3 metres**), this defines the vehicle on which the weight rests as 'the trailer'.

Regulations 139(d)(ii) and 140(4)(a) taken together thus make it clear that, where a load is carried by an Articulated Unit and one trailer, the former is to be excluded in calculating the 18.3 metres (60ft) limit.

Certain qualifications apply to this 'sixty-foot rule':-
(a) The rule will apply where the load is carried partly on a 'dolly' trailer and partly on a turntable superimposed on the drawing unit of an articulated combination (as this then becomes), and
(b) It will also apply where the load rests on more than one vehicle but is drawn by another vehicle on which it does not rest. In such a case the carrying vehicles plus the load will count towards the 'sixty-foot rule', but the entire vehicle combination will also have to be considered within the 'eighty-five foot rule'.

It is also relevant here to refer to the maximum overall length of an articulated combination where the semi-trailer is specially constructed and normally used for the carriage of indivisible loads of exceptional length (24). Where this overall length exceeds 16.8 metres (55ft), Police notification is required, but an attendant need not be supplied unless the overall length of the vehicle plus the load exceeds 18.3 metres (60ft). (See Fig. 11 - Overall Length).

7.2.2 Overall Width (Of the vehicle and its load)

There are absolute prohibitions relating to the maximum overall width of a vehicle and its load under both the C & U Regulations and the Special Types Order, as follows:-
(a) Under C & U Regulations section 140 the maximum width is 4.3 metres (14ft).
(b) Under Special Types Authorisation relating respectively to Abnormal Indivisible Loads and Engineering Plant, the maximum width is 6.1 metres (20ft), BUT the approval of the DTp is required where the width exceeds 4.3 metres (14ft).
(See Fig. 12).

7.2.3 Projecting Loads

A load may project forwards, rearwards or sideways (laterally) beyond the carrying vehicle(s).

Rearwards projections In measuring the amount of projection to the rear of a vehicle, account must be taken of the fact that the overall length of a vehicle

---

(24) Regulation 140(6)(e).

Fig. 10 LOAD DISTRIBUTION

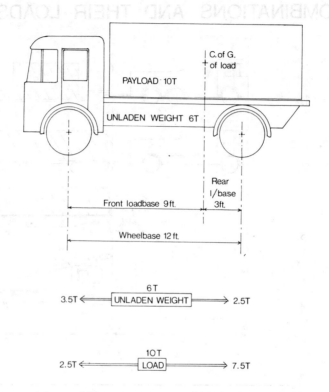

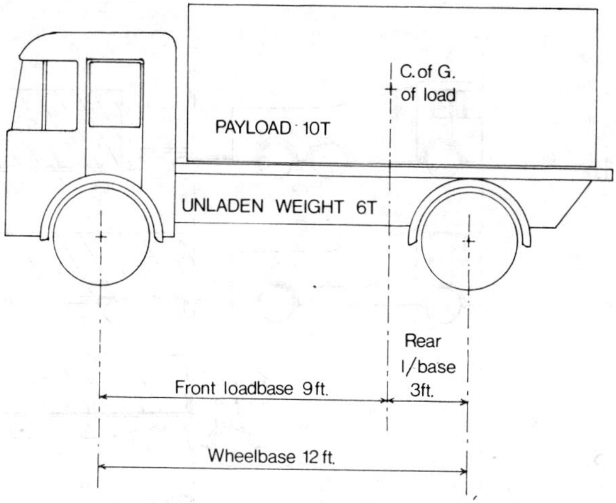

6T

3.5T ⟸ UNLADEN WEIGHT ⟹ 2.5T

10T

2.5T ⟸ LOAD ⟹ 7.5T

PLATED WEIGHTS:

16T

6T ⟸ GROSS VEHICLE WEIGHT ⟹ 10T

CALCULATION

Weight of load imposed on rear axle $= \dfrac{\text{Front Loadbase}}{\text{Wheelbase}} \times \text{Payload}$

$= 7.5T$

Weight of load imposed on front axle $=$ Payload $- 7.5T$

$= 2.5T$

Fig. 11 OVERALL LENGTHS OF VEHICLES, COMBINATIONS AND THEIR LOADS

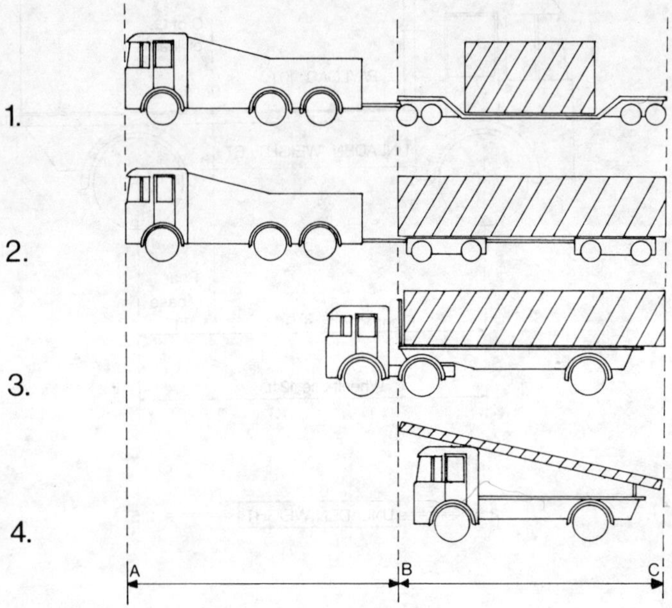

	OVERALL LENGTH OF LOAD AND SUPPORTING VEHICLES	OVERALL LENGTH OF COMBINATION	SUPPLY A MATE ?	2 DAYS NOTICE ?
1	(Under 60ft) (18.3 metres) *	A – C over 85ft (25.9 metres)	YES	YES
2	B – C over 60ft (18.3 metres)	(Under 85ft) (25.9 metres) *	YES	YES
3	B – C over 60ft (18.3 metres)	(Under 85ft) (25.9 metres) *	YES	YES
		Articulated combination over 55ft (16.8 metres) where the semi-trailer is specially constructed and normally used to carry loads of exceptional length (C & U Reg 140(6)(e)	NO	YES
4	B – C over 60ft (18.3 metres)		YES	YES

* – This information is given for completeness and has no direct bearing

FIG. 12
SIDEWAYS (LATERAL) PROJECTIONS

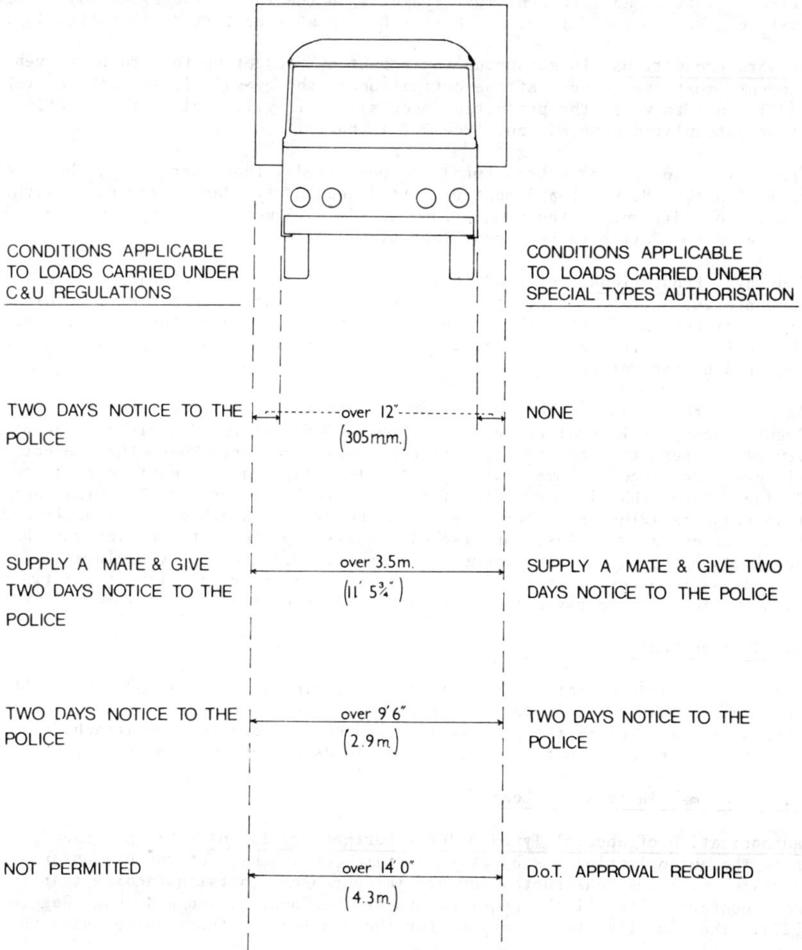

(additional to which the projection occurs) is exclusive of the tailboard (25).

Projections up to 1.07 metres (3ft 6ins) are permitted under both C & U Regulations and Special Types Orders, but beyond this length steps must be taken to make the load clearly visible, e.g. by tying a white rag to the load or splashing it with white paint. Where the projection exceeds 1.83 metres (6ft), there is a requirement to fit a Marker Board. Where the projection exceeds 3.05 metres (10ft), an attendant must travel with the load, two clear days' notice must be given to the Police, and Marker Boards must be fitted. (See Fig. 13).

Forward projections In measuring the amount of projection forward of a vehicle, account must be taken of the definition of the overall length of a vehicle (additional to which the projection occurs). This will include the tractive unit of an articulated combination. (See 7.2.1 above).

Projections up to 1.83 metres (6ft) are permitted without formality, but beyond that length a Marker Board must be fitted and an attendant must travel with the load. In addition, if the projection exceeds 3.05 metres (10ft), two clear days' notice to the Police is required. (See Fig. 13).

Sideways Projections (referred to in the regulations as 'lateral projections'). The provisions of the Construction and Use Regulations and the Special Types Orders differ markedly in this case. In both cases, where the overall width of the vehicle and its load exceeds 2.9 metres (9ft 6ins), two clear days' notice is required by the Police.

In all other cases where the load is carried under Construction and Use Regulations, a lateral projection of up to 305mm (12ins) on either side of the vehicle is permitted before any action becomes necessary. Where the projection at either side exceeds 305mm (12ins), two clear days' notice must be given to the Police. Note that if the width of the vehicle itself exceeds 7ft 6ins and the load projects 12ins on either side, then the overall width of the vehicle and its load will exceed 9ft 6ins, and two clear days' notice to the Police will have to be given on this count alone. Where the overall width of the vehicle and its load exceeds 3.5 metres (11ft 6ins), one attendant must travel with it and two clear days' notice must be given to the Police. (See Fig. 12).

7.2.4 High Loads

Where the laden weight of an articulated combination exceeds 32,520kg AND the semi-trailer has a plated weight exceeding 26,000 kg, a height restriction of 4.2 metres is applied to the semi-trailer and any rigid structure attached to the semi-trailer to contain the load. Sheeted loads are exempt from this restriction.

7.3 Abnormal Indivisible Loads

Authorisation of Special Types Order - Further Provisions (26) The Order provides for the authorisation, subject to certain conditions, of the use on roads of certain vehicles constructed for special purposes, notwithstanding that they do not conform with all the requirements of the Construction and Use Regulations (27). The most important sections for the haulier are those authorising the use

---

(25) Andrews v. Kershaw 1952. The tailboard is only counted in the overall length if the load actually rests on it, rather than merely passing over it when the tailboard is let down to accommodate the load.
(26) Motor Vehicles (Authorisation of Special Types) General Order, 1979.
(27) Ibid., Articles 20 and 21.

FIG. 13
FOREWARD & REARWARD PROJECTIONS

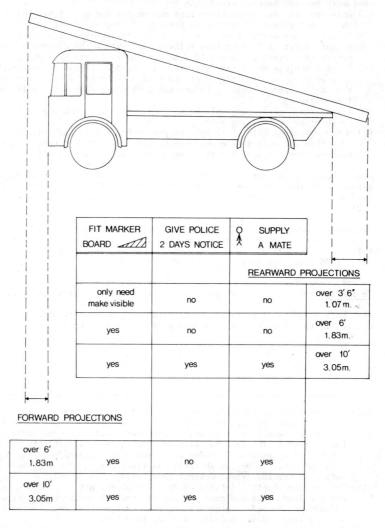

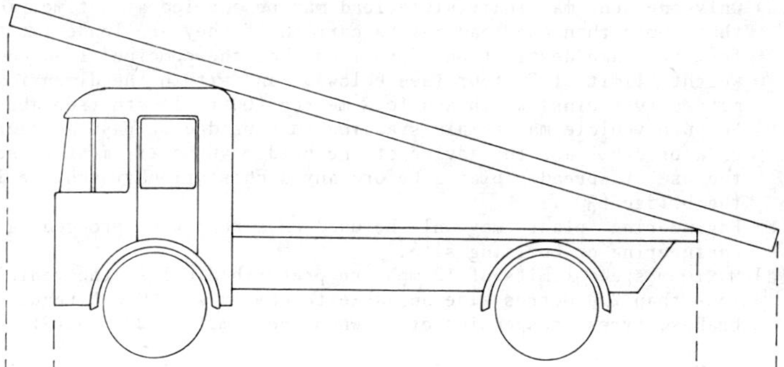

REARWARD PROJECTIONS

REARWARD PROJECTIONS	FIT MARKER BOARD	GIVE POLICE 2 DAYS NOTICE	SUPPLY A MATE
over 3′ 6″ 1.07 m.	only need make visible	no	no
over 6′ 1.83m.	yes	no	no
over 10′ 3.05m.	yes	yes	yes

FORWARD PROJECTIONS

FORWARD PROJECTIONS	FIT MARKER BOARD	GIVE POLICE 2 DAYS NOTICE	SUPPLY A MATE
over 6′ 1.83m	yes	no	yes
over 10′ 3.05m	yes	yes	yes

of vehicles for the carriage of abnormal indivisible loads (28) and engineering plant (29) respectively. The most essential conditions are that the vehicles used should be wheeled and fitted with pneumatic or soft elastic tyres, and that they must only be used as prescribed, e.g.:-

(a) Only one abnormal indivisible load may be carried at a time. The exception is that more than one load may be carried if they are loaded at the same place for the same destination, provided that the combined load comes within the weight limit of 75 tons (see below), and within the dimension limits of 2.9 metres (9ft 6ins) width and 18.3 metres (60ft) length (see above).
(b) No such vehicle may remain standing on a bridge or pass another such vehicle on a bridge, and the advice of the bridge authority must be sought regarding the use of spreader plates before any such stationary vehicle is jacked up on the bridge (30).
(c) Engineering plant may only be used on a road when proceeding to or from an engineering or building site.
(d) Maximum speed limits of 12 mph are prescribed unless the vehicle is unladen, less than 2.9 metres wide and able to comply with C & U regulations regarding brakes, tyres, suspension etc., when the limit is 20 mph (31).

Goods Vehicles operating under Special Types Regulations may display a plate marked with the operational weights which the manufacturer states should not be exceeded when the vehicle is operating loaded at speeds not in excess of 12 mph (32).

Generally, wherever the maximum Gross, Axle or Transmitted Weights specified in the Construction and Use Regulations are exceeded, two clear days' notice to the Police is required; but where the Gross Train Weight of a vehicle or combination exceeds 75 tons, six days' notice is required (33).

In all cases where attendants (34) are required, if three or more vehicles travel <u>in convoy</u>, the requirements apply only to the first and last vehicles.

The Police are also given powers to delay any movement of which they are given notice, in the interests of road safety and to prevent congestion.

---

(28) An abnormal indivisible load is defined as 'a load which cannot, without undue expense or risk of damage, be divided into two or more loads, and therefore cannot be carried in conformity with the Construction and Use Regulations'. The load may also be abnormally indivisible by reason of its weight if it exceeds 32,520kg or 32 tons Gross Plated Weight. A container, used as such, is <u>not</u> an abnormal indivisible load (Motor Vehicles (Authorisation of Special Types) General Order 1979, article 19).
(29) Engineering plant is defined as 'movable plant, not in itself constructed to carry a load, but designed and constructed for engineering purposes'. The definition extends to mobile cranes. (Motor Vehicles (Authorisation of Special Types) General Order 1979, article 19).
(30) Motor Vehicles (Authorisation of Special Types) General Order 1979, Articles 29 and 30.
(31) Ibid., Article 23.
(32) Road Vehicles (Marking of Special Weights) Regulation 1983 (S.I.910); Road Traffic Act 1972, section 42.
(33) Motor Vehicles (Authorisation of Special Types) General Order 1979, Articles 26-28.
(34) Ibid., Article 24.

8 Mechanical Condition

8.1 The Transport Manager's Responsibility

Since the owner-driver, or the 'person whose servant the driver is' (1) is the **user** of a goods vehicle, the Transport Manager by definition is either the user or his employee/agent (2). As such he is directly responsible for the mechanical condition of his vehicles.

8.1.1 Preventative Maintenance

The Department of Transport (3) are at pains not to lay down any yardsticks, because so much depends on the type of vehicle, the work it is doing and the mileage it is operating. They point out that the success (or otherwise) of the system employed must in the long run be reflected in the general condition of the fleet.

However, it is considered that regular **preventative maintenance** implies more than just carrying out recommended servicing. Systematic checks on items influencing the safety and roadworthiness of the vehicle should be carried out and recorded. A fair guide to items which require to be checked is listed in the Department of Transport's Goods Vehicle Tester's Manual (4). Competent staff should be employed to do this. It is up to the operator to choose whether inspections should be on a mileage or time basis, much depending on the use of the vehicle as suggested above. The important thing is that inspections must be seen to be systematic and adequate.

The question of facilities for inspection is also open to the judgement of the operator. Obviously, some vehicles can be inspected on a hard standing without the need for pits or ramps; equally obviously, others cannot. Owners of large fleets might find it essential to own certain types of equipment; smaller operators could find it sufficient to have access to these items.

Pre-inspection of a vehicle before first examination or periodic test at the Department of Transport Testing Station, together with the necessary preparation for the test (under-chassis cleaning etc.) can usefully be made to coincide with one of the vehicle inspections. Indeed, if vehicles are to be inspected on a time basis, this annual event could be made the key date around which the other inspections can be arranged.

---

(1) Transport Act 1968, section 92(2).
(2) The Goods Vehicle Operators (O-Licences, Qualifications and Fees) Regulations 1984 specify that the Transport Manager must be a full-time employee.
(3) Guide to Operators' Licensing - Part 8 - Maintenance Facilities.
(4) Published by HMSO.

The above is not, and was not intended to be, a treatise on the fleet engineer's responsibilities and problems; it merely seeks to paint guidelines on a very broad canvas. Many publications by trade associations, oil companies, manufacturers of equipment and transport journals provide a fund of useful detailed advice. Indeed, the operator is often inundated with this, and may feel the need to refer from time to time to general guidelines as above. The publication of maintenance check sheets and recording systems by the sorts of organisations mentioned above can be of real assistance in meeting the law's requirements relating to maintenance.

8.1.2 Responsibilities of the Operator

Even if an operator
(a) contracts out of the maintenance of his vehicles; or
(b) is the hirer of vehicles which the owner is contracted to maintain; or
(c) hires a trailer, or uses another operator's trailer,
he is still the legal 'user' of the vehicles, and responsible in law for the condition of those vehicles. He cannot contract out of this responsibility, and he must therefore still have an adequate system of inspection, and record all the maintenance done to the vehicles on his behalf. A difficulty arises where an operator hires a driver under a 'contract for service' (i.e. the driver is an independent contractor, not a servant of the operator under a 'contract of service' to him). In such a case, the employer of the driver becomes the user of the vehicle, and must hold the operator's licence and be responsible for the condition of the vehicle (5).

8.1.3 Defect Reports, Inspections and Records

The Road Traffic Act 1972 (6) imposes a duty on operators to inspect, and keep records of the inspection of, goods vehicles. Records of inspection must be retained for a minimum period of 15 months (7).

Form of the Records The minimum information which records must contain is:-
(a) Details of the place and date of any inspection carried out on a vehicle;
(b) Any defects discovered on the inspection;
(c) Details of any defects reported by the driver of the vehicle (a system must provide for the driver of each vehicle to report in writing any defects on the vehicle to the person responsible for rectifying them);
(d) Details of the date and place of any action taken to rectify any defects occurring as at (b) and (c) above.

The records must be available for inspection by a Department of Transport Examiner if called for during the 15 months period of retention.

8.2 Fleet Inspections and Roadside Checks

Apart from the first examination and periodic test of goods vehicles (Plating and Testing), Department of Transport Vehicle Examiners are empowered to inspect vehicles on the road, or to visit operators' premises, to ensure that vehicles are complying with the standards of safety and roadworthiness. These powers are contained in the Road Traffic Act 1972 (8), and relate only to goods vehicles.

---

(5) Ready Mixed Concrete (East Midlands) Ltd v. Yorkshire Traffic Area Licensing Authority, 1970.
(6) Road Traffic Act 1972, section 59.
(7) NB - If a driver's report form is incorporated in the driver's log book, it is important that when the latter is disposed of after 12 months, the former is retained for a further 3 months.
(8) Road Traffic Act 1972, section 56.

Another section of the same Act authorises Examiners and Police Officers (authorised in writing by a Chief Officer of Police) to inspect any motor vehicle or trailer on a road (9). In this section there is provision for the driver to elect for a deferred test, which may only be refused by a Police Officer if, after an accident, he considers the vehicle to be unsafe to proceed until tested. There is no provision for a deferred test of goods vehicles under section 56, and it is under this section that an Examiner will usually inspect goods vehicles. An extension of these powers of Examiners and Police Officers is contained in section 56 (10); either is empowered to direct the driver of a goods vehicle to proceed to a place not more than five miles distant (11), where a test can be carried out. Department of Transport Vehicle Examiners, on production of their authority, may also examine a goods vehicle at any reasonable time on the premises where it is known to be kept (12).

If the Examiner considers the vehicle to be unfit, he can issue a notice of prohibition of use on the roads (13) (form GV9). This may be either an 'immediate prohibition' or a 'deferred prohibition'; the latter specifies a date beyond which the vehicle may not be used unless the defect has been cleared following the examiner's re-inspection of the vehicle.

The GV9 relates to the use of the vehicle, and not merely to the carriage of goods (14). It will allow the user, when he claims to have remedied the defect, to take the vehicle to a testing station for clearance, or to road test it within 3 miles of the place where it has been repaired.

Form GV9A enables an examiner to vary the terms of any prohibition.

Form GV9B contains a number of clauses permitting the vehicle to proceed on certain conditions specified by the Examiner, who marks these on the form. For example, the vehicle may be allowed to travel (a) unladen, (b) below a specified maximum speed, (c) with a rigid or suspended tow, (d) outside 'lighting up' times, or (e) to a specified place.

Form GV9C constitutes a refusal to remove a prohibition.

Form GV10 (removal of a prohibition) is issued by the Examiner when a vehicle is presented with the defect cleared.

A GVDN (Defect Notice) is used to notify an operator of minor defects which do not warrant the issue of a full prohibition.

---

(9) Road Traffic Act 1972, section 53.
(10) Road Traffic Act 1972, section 56(4).
(11) The distance was increased to five miles under the Transport Act 1978.
(12) Road Traffic Act 1972, section 56(2)(b). Also, under Construction & Use Regulations 1978 section 145, an examiner, or a police officer in uniform, or a PSV Certifying Officer, may inspect any vehicles on premises, with the owner's consent, to ascertain that certain requirements of the Regulations are met. Except in the case of a vehicle which has been reported in an accident, the owner may require 48 hours' notice, but otherwise it is an offence to obstruct an examiner in the performance of his duty.
(13) Road Traffic Act 1972, section 58. Also, section 69 of the Transport Act 1968 allows the Licensing Authority to suspend, curtail or revoke a licence if a prohibition (GV9) has been imposed on a vehicle of which the licence holder was the owner, whether or not it was specified on the licence.
(14) Prior to the 1972 Act, a GV9 could not be issued to an empty vehicle.

8.3 Construction and Use Regulations

The Minister of Transport has powers under the Road Traffic Acts (15) to make regulations about 'the use of motor vehicles on the road, their construction and equipment'. In this context, the Acts are known as 'the enabling Acts'. The ensuing Regulations are (currently), the Construction and Use Regulations 1978, which consolidated the earlier (1973) Regulations and subsequent additions and amendments. Essentially the Construction and Use Regulations are a set of rules governing certain aspects of the construction of vehicles and their use on roads. They are in fact arranged in a number of Parts, the first of which are the Construction Regulations, the remainder the Use Regulations, as follows:-

Part I	Regulations 1-7 (Definitions)	) Construction Regulations
Part II	Regulations 8-79	)
Part III	Regulations 80-144	) Use Regulations
Parts IV & V	Regulations 145-154 (Testing and Inspection)	)
Part VI	Schedules	

The Construction Regulations (Part II) are of course largely the responsibility of the manufacturer; indeed, they may well become superfluous when Type Approval begins to operate. The following regulations, however, concern matters which are largely under the control of the operator.

8.3.1 Regulation 18 (Speedometers)

It is obligatory to fit every motor vehicle with a speedometer which is readily visible by the driver and which is accurate to within a margin of 10% above or below the actual speed. An exception applies to vehicles so constructed that they are incapable of exceeding 10 m.p.h., and vehicles fitted with recording equipment bearing a designated approval mark (16).

8.3.2 Regulation 23 (Mirrors)

Every goods vehicle, including dual purpose vehicles but excluding locomotives and tractors, must be fitted with two mirrors which give a view to the rear. One of these must be outside the vehicle on the off-side, and visible to the driver either through a side window or through the portion of the windscreen which is swept by the windscreen wiper; this external mirror must be capable of being adjusted by the driver from the driving seat. The other mirror may be internal.

8.3.3 Regulations 27 and 28 (Windscreen Wipers and Washers)

All motor vehicles must be fitted with at least one efficient automatic windscreen wiper, and with windscreen washers. There is an exception for vehicles with opening windscreens.

8.3.4 Regulation 29 (Horn OR Audible Warning Device)

Every motor vehicle must be fitted with an instrument capable of giving audible and sufficient warning of its presence (17).

---

(15) Road Traffic Act 1972, Part II, section 40.
(16) Motor Vehicles (Construction & Use) (Amendment) (No. 7) Regulations 1980.
(17) The sound emitted by the audible warning device fitted to a motor vehicle first used after 1st August 1973 must be uniform and continuous (Regulation 29(2)).

8.3.5 Regulations 19 and 104 (Fuel Tanks)

Fuel tanks must be maintained so as to be free from leaks and reasonably secured from damage (18).

8.3.6 Regulations 30, 31 and 114-118 (Silencers and Noise)

All vehicles driven by internal combustion engines must be fitted with silencers to reduce 'as far as is reasonable' the noise caused by the exhaust.

All new vehicles registered after 1st April 1970 must be constructed so as to meet the new EEC Directive on noise (19).

The Use Regulations relating to noise are concerned with nuisance to other road users, and they prohibit the use of a vehicle on a road if the vehicle is making excessive noise. The driver may be guilty of the offence if he unreasonably causes the noise when he could have avoided doing so, but it will be a defence if he can show that the noise was caused by the negligence of some other person, whose duty it was to service the vehicle so as not to be a nuisance. Likewise, if the operator can show that the noise could not reasonably have been prevented by due diligence (i.e. that it was accidental in nature), this will be a good defence. The 9th Schedule to the Regulations lays down a maximum noise level of 92 decibels for Plated Goods Vehicles (by definition those over 3.5 tonnes Gross Plated Weight), and 88 decibels for small goods vehicles and motor cars.

The same Schedule, as amended in 1980, lays down the following maximum noise levels for new motor vehicles first used in 1983.

Maximum Gross Weight of Vehicle	**Maximum sound levels (A Weighting) in decibels (dBA) measured when the vehicle is in use on the road**
Up to 3.5 tonnes	81dBA
3.5 - 12.0 tonnes	86dBA
Over 12.0 tonnes	88dBA

(Lower limits are prescribed where the sound level is measured under 'controlled conditions'.

It is an offence to leave a vehicle stationary (other than at an enforced traffic stop) or unattended with the engine running, unless the engine is required to drive some ancillary equipment through a power take off (PTO); or to use the horn while the vehicle is stationary or on a restricted road between 2330 and 0700 hours.

8.3.7 Regulations 33 and 34 (Emission of Smoke)

Regulation 33 prohibits the use of a vehicle on a road if the vehicle is emitting any smoke, visible vapour, grit, sparks, ashes, or oily substance in such a way as to be a nuisance to other road users. Pursuant to this, Regulation 34 prohibits the use of the excess fuel device on a diesel engine when the vehicle is in motion.

8.3.8 Regulations 107 and 108 (Tyres)

The regulations prohibit the use of a motor vehicle or trailer having a tyre, or tyres, which are:-

(a) incorrectly inflated;

---

(18) Fuel tanks on vehicles first used after 1st July 1973 must be made of metal.

(19) EEC Regulations 157/70 and 350/73.

(b) with a cut deep enough to reach the ply cords and over 25mm. long, or 10% of the section width of the tyre; or having a lump or bulge caused by separation or failure of the structure;
(c) with any portion of the ply cords exposed;
(d) without at least 1mm. tread depth in the tread pattern (but excluding any tie bar or tread wear indicator) throughout a continuous band measuring at least 75% of the breadth of the tyre (20);
(e) unsuitable for the use to which it is being put (e.g. insufficient ply rating, see Plating and Testing - Tyres) or incompatible with the other tyres on the vehicle (e.g. mixed radial and cross-ply tyres).

Recut tyres are illegal for all motor cars and some commercial vehicles.

8.3.9 Sideguards

The Regulations permitting the operation of five-axled articulated combinations at 38,000 kgs GPW (21) required that all trailers (22) and semi-trailers (23) with an unladen weight exceeding 1,020 kgs (1 ton) built on or after 1st May 1983, should be fitted with sideguards. Motor vehicles (22) over 3.5 tonnes GPW manufactured on or after 1st October 1983 and first used on or after 1st April 1984 are also required to be fitted with sideguards. These sideguards must satisfy dimensional and strength specifications in the Regulations.

Certain existing semi-trailers of whatever age, if used with a tractive unit whose GPW exceeds 32,520 kg (32 tons), must also be fitted with sideguards although these need only satisfy the dimensional requirements. The semi-trailers concerned are those whose GPW exceeds 26,000 kg, and the distance between the centre of the leading axle and the kingpin exceeds 4.5 metres. If used with a tractive unit over 32,520 kg, even if they are only tandem axle semi-trailers (which would limit the combination's GPW to 32,520 kg) they must be fitted with sideguards. Certain types of vehicles, including tippers, are exempt from the sideguards requirement.

8.3.10 Rear Under-run Protection

Rear under-run bumpers meeting both the dimensional and strength requirements of EEC Directives 221/1970 and 490/1979 must be fitted to all goods vehicles over 3.5 tonnes GPW and trailers and semi-trailers exceeding 1,020 kg unladen weight, if they were manufactured after 1st October 1983 and first used after 1st April 1984.

Certain vehicles including rear tippers are exempt, and alternative dimensional requirements may be met to accommodate tail lifts, demountable and other specialised bodies.

8.3.11 Ground Clearance

Trailers and semi-trailers manufactured after 1st April 1984 must now meet

(20) Or, where the original tread pattern did not extend beyond 75% of the breadth of the tread, the base of any groove in the original pattern is not at least 1mm deep. 'Breadth of tread' means that part of the tyre in contact with the road. (Motor Vehicles (Construction and Use) (Amendment) (No.3) Regulations 1983). Thus it becomes illegal for any tyre to show any signs of baldness where it is in contact with the road surface.
(21) Motor Vehicles (Construction and Use) (Amendment) (No.7) Regulations 1982 (S.I. No.1576).
(22) with an axle spread exceeding 3 metres.
(23) where the distance between the centre of the leading axle and the kingpin exceeds 4.5 metres.

minimum ground clearance specifications (24).

Trailer wheelbase or semi-trailer interspace (25)	Minimum Ground Clearance
6 metres and not exceeding 11.5 metres	160 mm
Exceeding 11.5 metres	190 mm

8.3.12 Regulations 126, 128, 137 and 138 (Towing and Attendants)

The driver must be readily able to operate any trailer brakes required to be fitted by the Construction and Use Regulations, unless these are of the over-run or self-operating type. If the distance between the towing vehicle and trailer is more than 1.5 metres, the chain or rope used must be made readily visible from the side.

The position regarding the employment of attendants where trailers are drawn is as follows.
(1) The number of trailers which can be drawn is limited as follows (26):-

Type of Towing Vehicle	Number of trailers which can be drawn	
	Laden	Unladen
Heavy or light locomotive	3	3
Tractor	1	2
Motor Car or Heavy Motor Car	1	1

Notes:-
(a) A broken-down vehicle being towed, or an **unladen** articulated vehicle being towed, count as one trailer (27).
(b) A towing implement on which rests, or from which is suspended, another vehicle, together with the implement itself being towed, count as one trailer (28).
(c) A composite trailer (i.e. a combination of a semi-trailer and a converter dolly (29)) is treated as a single trailer for the purpose of Regulation 137.

(2) The Road Traffic Act 1972 (30) prescribes the number of attendants required in the above instances, as being two attendants in the case of a locomotive, plus one additional attendant for each trailer drawn in excess of one, and one attendant in the case of a tractor, motor car or heavy motor car drawing a trailer or

---

(24) Motor Vehicles (Construction and Use) (Amendment) (No. 2) Regulations 1983.
(25) Interspace in this context is the distance between the rear axle (or the centre of the rear bogie) and the front axle or kingpin as the case may be.
(26) Road Traffic Act 1974, schedule 6, paragraph 16; Motor Vehicles (Construction and Use) Regulations 1978, section 137.
(27) Road Traffic Act 1972, section 65; Motor Vehicles (Construction and Use) Regulations 1978, section 137.
(28) Motor Vehicles (Construction and Use) Regulations 1978, section 137.
(29) A 'converter dolly' is a trailer with two or more wheels designed to enable a semi-trailer to move without any part of its weight resting on the drawing vehicle (Construction and Use Regulations 1978, section 3(1)). Both the dolly and the semi-trailer must be plated and maintained so that the outfit achieves post '68 braking standards (50%/25%/16%).
(30) Road Traffic Act 1972, section 34.

trailers.

(3) The Construction and Use Regulations (31), however, make a number of very important exceptions. No attendant is required in the case of:-

(a) an articulated vehicle;
(b) a solo locomotive propelled by liquid fuel or electrical power;
(c) a motor vehicle drawing a trailer fitted with power or power-assisted brakes which can be operated by the driver. (If more than one trailer is drawn, e.g. by a tractor, then only one attendant will be required).

The above effectively permits the operation of rigid vehicles with drawbar trailers without attendants (mates).

8.3.13 Braking (C & U Regulations 13, 14, 51, 55, 59 and 64)

Each section of the Construction and Use Regulations (32) relating to each particular class of vehicle makes reference to the Braking Standards required. As far as goods vehicles are concerned, the regulations can be summarised in table form as follows:-

CLASS OF VEHICLE	EFFICIENCY REQUIRED Main Brake	Secondary Brake	Parking Brake
Vehicles registered after 1.1.68	50%	25%	16%
Vehicles registered before 1.1.68 - with 2 axles	45%	20%	-
Vehicles registered before 1.1.68 - with 3 or more axles (including artics)	40%	15%	-

Ref:- Motor Vehicles (Construction and Use) Regulations 1978.

The regulations applicable to vehicles registered before 1.1.68 are sometimes referred to as the Interim Braking Efficiencies, as it is intended that these lower standards should eventually be withdrawn.

8.3.13.1 Braking Efficiency

Braking Efficiency is defined quite simply as $\frac{\text{Braking Force}}{\text{Gross Vehicle Weight}}$.

The braking force of a vehicle can be measured on a dynamometer or 'rolling road', and by dividing this figure (which will be a constant at all weights) by the GVW, the Braking Efficiency is obtained. Obviously, the higher the GVW, the

(31) Regulation 138(1)(a) in the case of articulated vehicles; 138(2) in the case of locomotives; 138(1)(m) in the case of drawbar trailers.

(32) Motor Vehicles (Construction and Use) Regulations 1978, sections 13, 14, 51, 55, 59, 64 and schedule 4; EEC Directives 320/71, 132/74 and 489/79.

lower the Braking Efficiency. The Braking Efficiency figures above have to be met at the Plated Gross Weight of the vehicle. In theory, if the vehicle had only a manufacturer's plate with a design weight greater than the maximum permitted C & U weight (i.e. the weight at which the DTp would 'plate' the vehicle), then it would be required, for instance at a roadside check, to meet its required braking efficiency at that design weight. However, in a letter to the Freight Transport Association (then the TRTA) in May 1968, the DTp gave a declaration of intent that in such circumstances braking performance would be tested at C & U permitted weights (33). There is also a relationship between braking efficiency and the gradient of a slope upon which the brakes of the vehicle will hold. Regulation 13 prescribes a braking efficiency of 16% for parking brakes of new vehicles by saying that they must be capable of holding the vehicle on an incline of 1 in 6.25. It can be shown mathematically that if this happens, the Braking Force of the vehicle and its Gross Vehicle Weight must be in the proportion 1:6.25, i.e.

$$\frac{\text{Braking Force}}{\text{GVW}} = \frac{1}{6.25} \times 100\% = 16\%.$$

If for example the dynamometer registered a total Braking Force of 1,000 kg on all wheels, and if the GVW was 6,250 kg, the brakes would be 16% efficient. By the same token, if the vehicle's brakes held it on the 1 in 6.25 incline which is constructed outside DTp Test Stations, its brakes would be at least 16% efficient. Incidentally, the maximum gradient of motorway entrance/exit ramps is 1 in 6.25.

8.3.13.2 Types of Brakes

At this point it is relevant to consider and define the types of braking systems referred to in the Table on the previous page. The concept of a vehicle having a footbrake and handbrake is somewhat imprecise, and the regulations relate to Main, Secondary and Parking Brakes.

As will be seen, almost invariably the footbrake will be the Main Brake of the vehicle. The handbrake may be either a Secondary Brake, or a Parking Brake, or both!

The Main Brake of a vehicle is defined as the main means normally available to the driver of bringing the vehicle to rest. Sometimes it is referred to as the 'service brake'. The definition needs no comment.

The Secondary Brake is defined as a braking system which, notwithstanding the failure of any part of the main system, is available for application by the driver to at least half the number of roadwheels. This is often referred to as a 'Fail-Safe' braking system. The requirement that it should work on half the number of roadwheels goes a long way towards explaining why the required secondary braking efficiency is in the region of half the main braking efficiency. There is a further requirement that the secondary braking system should be a mechanical system unless the requirement regarding parking brake standards is already met by a mechanical system. This proviso in fact allows a number of systems to qualify as a secondary brake, including split line air systems, 'dead man's handles', and three-line air braking systems on articulated combinations.

The Parking Brake is defined as a means of holding the vehicle stationary. As stated earlier, on 'new' vehicles it must do so with the vehicle on a 1 in 6.25 incline. A spring brake which operates by means of air pressure holding the brake

---

(33) 'Motor Transport', 3rd May 1968. There will be very few such vehicles now that the maximum Gross Plated Weight under the C & U Regulations is 38,000 kgs.

off, and applying the brake when released, will qualify as a mechanical brake. Such a system is definitely 'fail-safe'; in fact, it is suggested that one reason for the third lane ban on goods vehicles on motorways might be the widespread use of such braking systems, and the possible consequences of their 'failing' in a fast lane.

Regulation 101(2), (3) and (4), although really a 'Use' regulation inasmuch as it prohibits the use of a motor vehicle with a trailer or semi-trailer so combined as to produce an overall braking efficiency of less than the required standard, has important construction connotations. In effect it means that although the motor vehicle when used solo must have at least two independent braking systems (as must all motor vehicles), when the motor vehicle is coupled with a trailer its own braking system, which may have been quite adequate at the motor vehicle's own GVW, may well be inadequate at the Gross Train Weight or Gross Combination Weight. In such circumstances the braking force of the trailer brakes may be taken into account, together with the braking force of the motor vehicle, in achieving the required efficiency. However, a trailer brake can never be a substitute for a secondary brake on the drawing vehicle, as this would leave the drawing vehicle with only one braking system when running solo.

Some vehicles are fitted with weight-sensing devices which effectively reduce the vehicles' braking force when the GVW is reduced, e.g. when running empty or lightly loaded. There are good safety reasons for this, particularly with articulated vehicles, because the system can reduce the incidence of wheel lock and skidding, itself a contributor to artic 'jack-knifing'. So long as the braking force is only reduced in the same proportion as the load, the system is quite legal, as the proportion of braking force to GVW can be kept at or above the required percentage efficiency.

8.3.13.3 EEC Braking Standards

Vehicles and trailers first manufactured during or after October 1982 and first used during or after April 1983 must be fitted with braking systems complying with EEC braking directives.

Under UK regulations the main and secondary braking efficiencies can be met by a dual braking system sharing the same control (each half of the system achieving 25% efficiency - total 50%) but in this instance, under EEC regulations, the <u>parking brake</u> must then be operable whilst the vehicle is in motion.

It must also be capable of holding the vehicle and any attached <u>unbraked</u> trailer on an incline of 12% (1 in 8.3).

On small Goods Vehicles under 3.5 tonnes GPW, if part of the service brake fails the residual efficiency must be 30%.

Spring brakes must be compressed via an independent energy supply, although being <u>mechanically operated</u> (by a spring) to comply with parking brake regulations. It must be possible to release them using a device (spanner) which must be carried on the vehicle.

8.3.14 Miscellaneous provisions of the Use Regulations (Part III)

The person responsible for ensuring that a vehicle is used in compliance with the Construction and Use Regulations is the user of the vehicle (1). Thus an operator (and perhaps the driver as well) may find himself prosecuted for an offence of which he was quite ignorant and which may have occurred many miles from his

---

(1) Transport Act 1968, section 92(2).

operating centre (34).

A vehicle is still being used even when stationary for loading purposes (35), and a driver may himself use a vehicle with an insecure load, contrary to the Construction and Use Regulations (36), even if someone else loaded the vehicle (37).

If the overall height of a vehicle, including its trailer or semi-trailer and load (where this is a container, engineering plant or skip hoist) exceeds 12ft 0ins, a notice indicating its height in feet and inches, and in letters at least 40mm. tall, must be fitted where the driver can see it.

Regulation 17 (Seat Belts and Anchorage Points), makes it obligatory to fit seat belts and properly designed anchorage points to the driver's and front passenger's seats of every motor car registered on or after 1st January 1965, but an exemption releases Goods Vehicles of over 1525 kg unladen weight (38) from this obligation. This exemption for goods vehicles cannot be applied to dual purpose vehicles, or passenger vehicles adapted to carry less than 12 passengers excluding the driver. Vehicles first used after 1st April 1973 must be fitted with seat belts which can be fitted and adjusted with one hand.

Compulsory use of Seat Belts The Transport Act 1981 empowers the Secretary of State for Transport to require those driving or riding in motor vehicles where the fitting of belts is now compulsory, to wear these (39). Since January 1983 seat belts, where required to be fitted, have to be worn by drivers and front seat passengers.

The Secretary of State may prescribe exemptions which must include
(a) drivers on local delivery rounds;
(b) drivers when manoeuvring or reversing;
(c) drivers concerning whom a doctor has issued a certificate that it would be unwise to wear belts.

There will be a maximum penalty of £50, but drivers will not be held responsible for the omissions of their adult passengers, and vice versa.

Children under the age of 14 may not travel in the front seat of a vehicle unless properly restrained.

Regulation 120 prohibits the reversing of a vehicle for longer than is reasonably necessary.

8.4 Lighting and Marking

The principal legislation concerning the lighting and marking of road vehicles is contained in the Road Vehicles Lighting Regulations 1984 (40), as well as parts of the Construction and Use Regulations.

---

(34) Green v. Burnett, 1955 - but see also Vehicle Insurance, defence under Road Traffic Act 1972.
(35) Andrews v. Kershaw, 1952.
(36) Construction and Use Regulations 1978, section 97.
(37) Gifford v. Whittacker, 1942.
(38) Or 3.5 tonnes if made after 1st October 1979 and first used after 1st April 1980 (Construction and Use Regulations section 17(12)).
(39) The Regulations must be approved by both Houses of Parliament, and will lapse in three years unless re-approved by both Houses.
(40) The Regulations incorporate EEC Directives 256/1976 and 933/1978, as amended in 1980, 1982, 1983 and 1984.

8.4.1 Statutory Lights

Sidelamps and Rear Lamps Sidelamps and rear lamps are also referred to as obligatory front and rear position lamps. Under the 1984 Regulations (41) they must be fitted to all vehicles used on the road during the hours of darkness. Hours of darkness (sometimes called 'lighting-up time') is defined as the period from half an hour after sunset until half an hour before sunrise. During this time the lamps must be illuminated.

Sidelamps Two lamps must be fitted, on opposite sides of the vehicle, at not more than 1.5 metres above the ground. On vehicles first used on or after 1st April 1986, and on trailers manufactured on or after 1st October 1985, the sidelamps must comply with the horizontal angles of visibility shown in Fig.17, and must be visible also from positions 15 degrees above and below the horizontal (42).

Rear lamps Two lamps must be fitted, on opposite sides of the vehicle, and at least 500mm apart. The lamps must either be visible from a reasonable distance, or bear an approval mark. They must be positioned so that they are
(a) not less than 350mm above the ground;
(b) not more than 2.1 metres above the ground (43);
(c) at or near the rearmost point of the vehicle.

If a vehicle is stationary on a road during the hours of darkness, other than for an enforced traffic stop, all lights other than the above obligatory lights, and any light not being red, illuminating the rear number plate, must be extinguished. From this it will be seen that the obligatory lights above are required to be illuminated on a vehicle **parked** on a road during the hours of darkness; but there are minor exceptions to this rule which allow private cars and goods vehicles under 1525 kg unladen weight to park with a single **parking light** or without any lights at all, if they are:-
(a) not within 10 metres of a junction;
(b) within an area where a speed limit of 30 m.p.h. or less applies; and
(c) only on the nearside of the road (44).

Headlamps The Road Traffic Act 1972, in empowering the Secretary of State to make headlamps obligatory for certain vehicles, allowed at the same time the combination of headlamps with the (already obligatory) sidelamps in a single unit (45).

Where the fitting of headlamps is obligatory, a minimum wattage of 30 watts is prescribed (46).

All vehicles with three or more wheels are required to be fitted with two headlamps capable of emitting a main beam and a dipped beam, or two groups of such headlamps so arranged that the outside lamps can emit a dipped beam and the remainder can emit main beams. The headlamps must be so wired that the operation of the vehicle's 'dip switch' will simultaneously extinguish every main beam and

---

(41) Road Vehicles Lighting Regulations 1984, section 16, and schedules 2 and 10.
(42) Road Vehicles Lighting Regulations 1984, Schedule 2(I).
(43) Motor Vehicles first used before 1st April 1986, and other vehicles manufactured before 1st October 1985, are no longer subject to the more stringent provisions of earlier legislation.
(44) Road Vehicles Lighting Regulations 1984, section 22.
(45) Road Vehicles Lighting Regulations 1984, sections 16 and 20. Vehicles must still have **both** side and head lamps at the front (Payne v. Holland, 1980).
(46) There is no minimum wattage requirement for Motor Vehicles first used on or after 1st April 1986.

Fig.14 POSITIONING OF HEADLAMPS

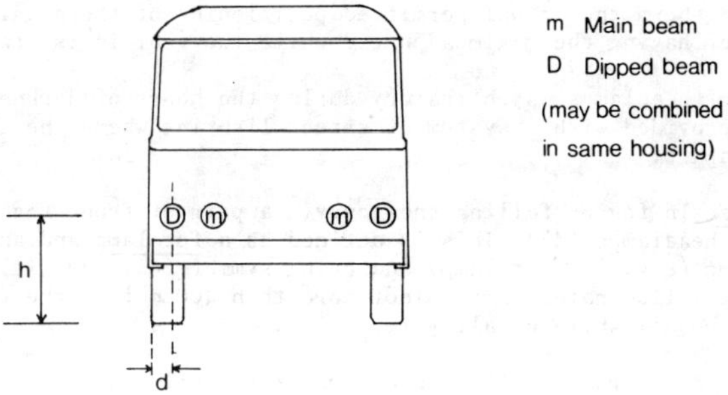

MAX DISTANCE FROM SIDE OF VEHICLE : d

400mm

HEIGHT : h

Maximum : 1200mm

Minimum : 500mm   Does not apply to Fog Lamps.

(Road Vehicle Lighting Regulations 1971)

operate every dipped beam. Each set of beams (main and dipped) must be capable of being switched on or off in pairs.

The headlamps must be matched to show either all yellow or all white light, and must be mounted symmetrically so that each pair of dipped beam headlamps is
(a) the same height above the ground, and
(b) the same horizontal disposition, and
(c) the same shape and area.
Figure 14 shows the actual permitted positioning of the headlamps. They must be capable of having their aim adjusted while the vehicle is stationary.

The use of headlamps is obligatory during the hours of darkness on any road which is not provided with a system of street lighting where the lights are 200 yards or less apart.

Fog Lamps In fog or falling snow only, a pair of front fog lamps may be used in place of headlamps (47). This is defined as a **fog lamp** and another lamp not being a headlamp (e.g. a spot lamp) and being symmetrically positioned, so that their outermost illuminated part is not more than 400mm from the vehicle's outer edge. They may be all white or all yellow.

Vehicles and trailers first used from 1st April 1980 must be fitted with rear **high intensity fog lamps** (48) - either a single lamp on or to the offside of the vehicle, or a matched pair, of the same type, at the same height above the ground, and equidistant from the centre line of the vehicle. They may be combined with other lamps in a common housing, and must be wired so that they are not illuminated by the application of the brakes (48). Their use must be indicated to the driver by a dashboard warning light.

Maintenance of Lighting Equipment Even if a vehicle is not used during the hours of darkness, it is still necessary for it to be fitted with obligatory lamps and with reflectors, except as provided below.

If the lamps are painted out or masked so that they cannot be readily used, or if they are not wired up, they need not be maintained in working order. In all other cases they must be maintained in working order, both during the hours of darkness and in daylight (49).

Defences It will generally be a good defence for the driver of a vehicle with defective lights to show either that:-
(a) The defect arose through the neglect or default of some other person, whose duty it was to provide the vehicle with lamps or reflectors (**BUT** this defence would not prevent proceedings being taken against any person causing or permitting the offence) (50); or that
(b) The defect occurred during the course of a journey, or that steps had already been taken to remedy the defect (51).
There is no statutory defence against the use of a motor vehicle with defective lights during the hours of darkness.

---

(47) Prior to 1st January 1970, fog lamps (which are not themselves obligatory) could be used only with headlamps. Dipped headlamps are also obligatory in conditions of poor visibility, e.g. heavy rain (Road Vehicles Lighting Regulations 1984, section 22(2)(b)).
(48) Road Vehicles Lighting Regulations 1984, section 16 and schedule 11.
(49) Road Vehicles Lighting Regulations 1984, section 4(4).
(50) Road Traffic Act 1972, section 81.
(51) Road Vehicles Lighting Regulations 1984, section 20(4)(c).

Fig.15

POSITIONING OF AMBER SIDE-FACING REFLECTORS

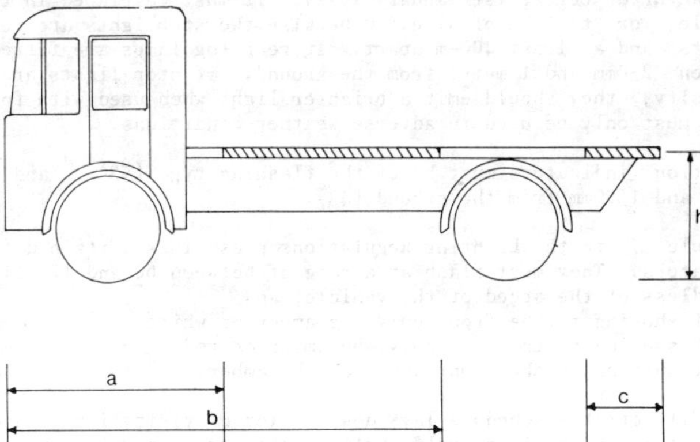

SIDE-FACING REFLECTORS WITHIN SHADED AREA

TABLE OF DIMENSIONS

	Motor Vehicle first used on or after 1st April 1986, or Trailer manufactured on or after 1st October 1985	All other Motor Vehicles and Trailers
a	3 metres maximum	$\frac{1}{3}$ length of vehicle
b	None more than 3 metres apart	$\frac{2}{3}$ length of vehicle
c	1 metre	1 metre
h	900 mm (or 1500 mm if the vehicle structure makes 900 mm impossible)	Minimum 350 mm, maximum 1500 mm

8.4.2 Stop Lamps and Direction Indicators

These are compulsory and must be maintained clean and in good working order (52). All stop lamps fitted to a vehicle and a trailer must show a red diffused light, of between 15-36 watts intensity.

Where only one stop lamp is required (on vehicles first used and on trailers manufactured before 1st January 1971), it must be placed in the centre of the vehicle, or to the offside. Otherwise two stoplights are required, at equal heights and at least 400mm apart. If rear fog lamps are fitted, they must be between 250mm and 1 metre from the ground. If stop lights are of optional dual intensity, they should emit a brighter light when used with fog lamps. Rear fog lamps must only be used in adverse weather conditions.

Direction indicators must be of the flashing type (53), and positioned between 350mm and 1500mm from the ground (43).

Schedule 7 to the Lighting Regulations prescribes certain details relating to indicators. They must flash at a rate of between 60 and 120 flashes per minute, regardless of the speed of the vehicle, and
(a) if showing to the front only, be amber or white;
(b) if showing to the rear only, be amber or red;
(c) if showing to the front and rear, be amber.

Part II of the schedule lays down angles of visibility. Incorporated in the schedule is an excellent self-explicit diagram, to which operators' attention is drawn. A particular point in the Regulations is the requirement for articulated combinations to be fitted with flank indicators.

From 1st July 1974, motor vehicles first registered and trailers manufactured after that date could be fitted with dual intensity stop lights and indicators. These should show a brighter light if operated while the vehicle's obligatory front and rear lights are switched off, than they show when the obligatory lights are in use. However, dual intensity indicators are not a requirement of the Road Vehicles Lighting Regulations 1984 (54).

8.4.3 Hazard Warning Devices

Vehicles first used on or after 1st April 1986 must be fitted with a hazard warning signal device, and a driver's 'tell-tale' to indicate their operation (55).

8.4.4 Reflectors

All motor vehicles used on a road, whether or not during the hours of darkness, must have fitted to the rear two red reflectors, not more than 400mm (56) from the edge of the vehicle and between 350mm (43) and 1.2 metres (57) above the

---

(43) Motor Vehicles first used before 1st April 1986, and other vehicles manufactured before 1st October 1985, are no longer subject to the more stringent provisions of earlier legislation.
(52) Road Vehicles Lighting Regulations 1984, section 20(1).
(53) Vehicles registered before September 1975 may have 'semaphore' indicators.
(54) If the stop lights are of optional dual intensity, they should also emit a brighter light when used with fog lamps.
(55) Road Vehicles Lighting Regulations 1984, schedule 8.
(56) 610mm for vehicles first used before 1st April 1986 and trailers manufactured before 1st October 1985.
(57) 1525mm for vehicles first used before 1st April 1986 and trailers manufactured before 1st October 1985.

ground. Because of the similarity of positioning of rear lamps and reflectors it is usually possible, and is permitted, to combine these.

All trailers and semi-trailers must be fitted with triangular rear reflectors (58).

Motor vehicles over 8 metres (59), and trailers over 5 metres in length, must be fitted with two side-facing amber reflectors on each side of the vehicle, at the rear and at the centre of the vehicle (See Figure 15). For Motor Vehicles first used on or after 1st April 1986, and trailers manufactured on or after 1st October 1985, these must be
(a) not more than 3 metres from the front;
(b) not more than 1 metre from the rear; and
(c) not more than 3 metres apart (60).

8.4.5 Side and Front Marker Lamps

Under certain conditions during the hours of darkness, additional lamps must be fitted and illuminated. The conditions apply to vehicles, vehicle combinations and their loads, irrespective of whether they are operating under the Construction and Use Regulations or a Special Types Order.

Side Marker Lamps These show a white light to the front, and a red light to the rear, through an arc of 70 degrees forward and rearward respectively of a line drawn through the lamp at right-angles to the longitudinal axis of the vehicle.

Where the overall length of a vehicle or vehicle combination, including the load, exceeds 18.3 metres, side marker lamps must be fitted at each side of the vehicle as follows (61) (See Figure 16):-
(a) One lamp in such a position that no part of its illuminated area is more than 9.15 metres from the foremost part of the vehicle(s) and load;
(b) One lamp in such a position that no part of its illuminated area is more than 3.05 metres from the rearmost part of the vehicle(s) and load;
(c) Such other lamps as will ensure that there is not more than a 3.05 metres gap between themselves and the lamps positioned as in (a) or (b) above.

The above conditions need not apply to a broken-down vehicle being towed, or a vehicle carrying a projecting load with an illuminated Marker Board.

Where a load is carried on, and supported by, two or more vehicles (not being an articulated combination) and the overall length of the combination is over 12.2 metres and under 18.3 metres, special conditions apply:-
(a) One lamp must be not forward of the extreme rear of the motor vehicle but within 1.53 metres to the rear of this point;
(b) If the load itself extends more than 9.15 metres behind the rear of the motor vehicle, there must be a lamp not in front of, and not more than 1.53 metres behind, the load's centre point.
(See Fig. 16). These conditions naturally apply to both sides of the vehicle.

If a drawbar trailer, or a semi-trailer, exceeds 9.15 metres in length, there must be a side marker lamp not in front of, and not more than 1.53 metres behind, the centre point of the trailer, on each side. (See Fig. 17).

---

(58) Complying with British Standard BSAU40.
(59) 6 metres for vehicles first used on or after 1st April 1986. Road Vehicles Lighting Regulations 1984, schedule 16, part I.
(60) Road Vehicles Lighting Regulations 1984, schedule 16, part I.
(61) Road Vehicles Lighting Regulations 1984, section 19, schedule 9.

Fig.16 FITTING OF SIDE MARKER LAMPS

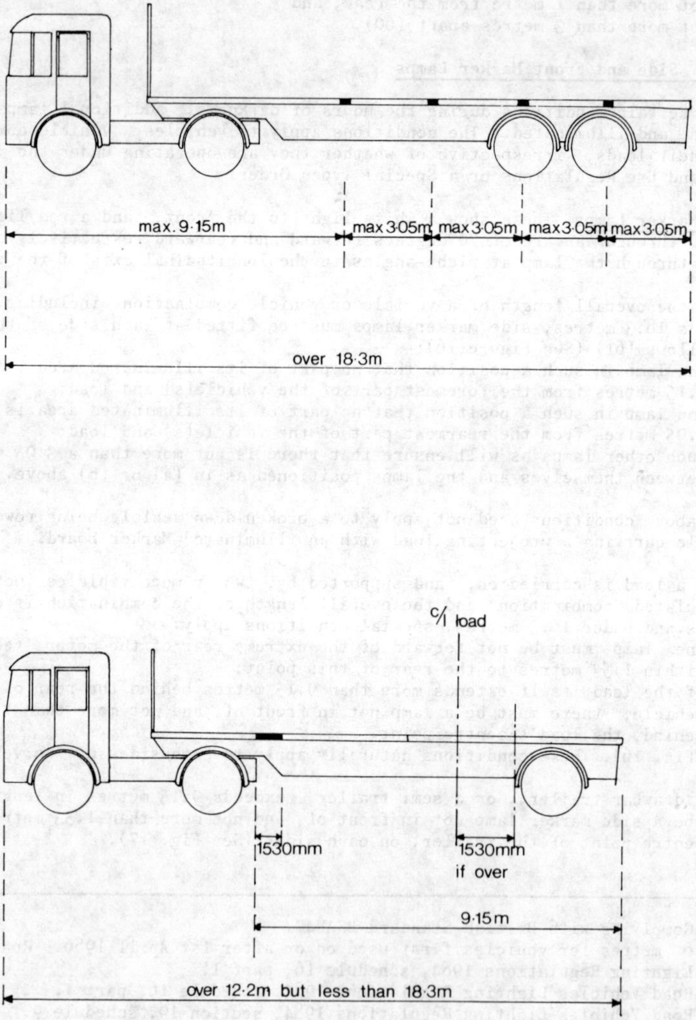

Fig.17 FITTING OF SIDE AND FRONT MARKER LAMPS TO TRAILERS AND SEMI-TRAILERS

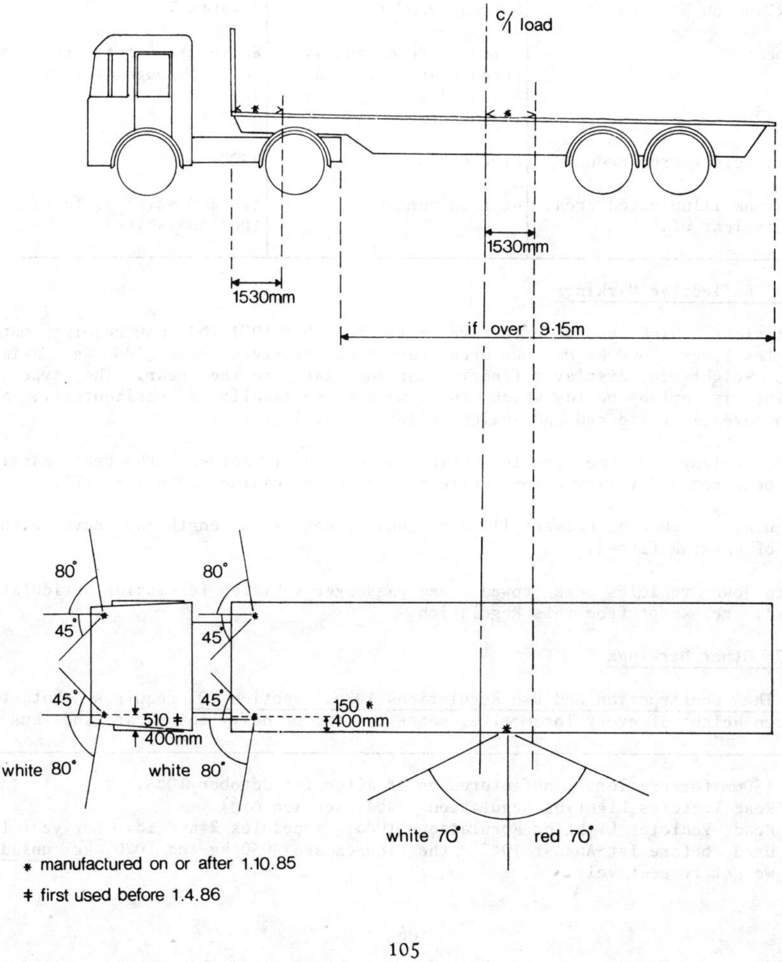

* manufactured on or after 1.10.85

ǂ first used before 1.4.86

Front Position Lamps These are often referred to as Front Corner Marker Lamps when fitted to trailers and semi-trailers. They may be positioned up to 400mm (62) inwards of the side of the vehicle.

In general, front corner marker lamps will be required on trailers and semi-trailers, but there are exceptions to this rule, and front corner marker lamps are not required in the following circumstances:-

(i) Where the trailer is a broken-down vehicle being towed (63);

(ii) Where the overall width of the trailer does not exceed 1.6 metres.

(iii) Where the trailer was manufactured before 1st October 1985 and its overall length, excluding any drawbar, does not exceed 2.3 metres.

The following additional requirements apply.

Requirement	Front Corner Marker Lamps	Side Marker Lamps
Maximum bulb wattage	No requirement	7 watts
Aspect	White to front and side through arcs of 80 deg. and 45 deg. respectively (See Figure 17)	White to front, red to rear through arcs of 70 deg.
Max. height from ground	2100 mm	2300 mm
Minimum illuminated area, equivalent of:-	No requirement	two 490 sq.mm surfaces (red and white)

8.4.6 Reflective Markings

Regulations which became effective on 1st November 1971 (64) now require motor vehicles over 7500 kg Maximum Gross Weight and trailers over 3500 kg Maximum Gross Weight to display reflective marking plates to the rear. The type of marking is oblong boards which can be mounted vertically or horizontally, and which have alternate red and yellow reflective stripes.

Where the length of the vehicle combination exceeds 13 metres, the rear marking must be a red and yellow oblong reflective marking reading 'LONG VEHICLE'.

Articulated vehicles between 11 metres and 13 metres in length may have either type of marking fitted.

Broken-down vehicles being towed, and passenger vehicles (excluding articulated buses), are exempt from this Regulation.

8.4.7 Other Markings

(1) The Construction and Use Regulations 1978, section 80, requires that the Unladen Weight of every locomotive, motor tractor or heavy motor car (and thus of

---

(62) 150mm for trailers manufactured on or after 1st October 1985.

(63) Road Vehicles Lighting Regulations 1984, section 6(6).

(64) Road Vehicles Lighting Regulations 1984, schedules 2 and 18. For vehicles used before 1st August 1982, the figures are 3050 kg and 1020 kg **unladen** weight respectively.

all Goods Vehicles weighing over 3 tons unladen) must be plainly marked in a conspicuous place on the nearside of the vehicle. The weight of the tractive unit of an articulated vehicle must be shown (65).

(2) **The Food Hygiene Regulations 1966** require that every vehicle used in connection with a food business must show conspicuously the name and address of the person carrying on the business.

(3) **The Weights and Measures Acts** (66) require that every vehicle of a capacity of more than 1 cubic yard, which is used for the carriage of sand and ballast which is to be sold by volume, must show on the outside of the body the maximum capacity, and must also bear the Weights and Measures Inspector's verification stamp.

(4) Vehicles which carry corrosive, inflammable or radioactive loads must carry prescribed symbols (67). In addition, where dangerous substances are carried in road tankers and road containers, full information on the nature of the load and risks involved must be given by the consignor to the carrier, the driver must be given full details of the load in writing (68), including recommended emergency action, and a training programme for all employees involved must be in force.

All road tankers, and any tank container with a capacity of more than 3 cubic metres, must display hazard warning (HAZCHEM) panels (69). See Fig. 18.

(5) Every trailer drawn by a motor vehicle must exhibit a **trailer plate** to the rear, unless special reflectors are used. A description of the trailer plate, which is a hollow white triangle studded with red reflex lenses, is given in Schedule 5 to the Construction and Use Regulations 1978. There is no need to display a trailer plate at the back of a semi-trailer, or of a broken-down vehicle being towed (70).

---

(65) However, this is no longer a requirement where a Ministry Plate has been issued which shows the vehicle's unladen weight. (Motor Vehicles (Type Approval for Goods Vehicles) (Great Britain) Regulations 1982.

(66) Weights and Measures Acts 1936 and 1963. The latter Act introduced metric equivalent measures in preparation for the metrication of road freight transport on 1st January 1974.

(67) The Inflammable Substances (Conveyance by Road) (Labelling) Regulations 1971, and the Dangerous Substances (Conveyance by Road in Road Tankers and Tank Containers) Regulations 1981, require that certain vehicles carrying inflammable substances and corrosive materials should display at the front and rear of the vehicle a label with the word INFLAMMABLE/CORROSIVE as appropriate. See also The Organic Peroxide (Conveyance by Road) Regulations 1973; The Radioactive Substance (Carriage by Road) (GB) Regulations 1974 and The Radioactive Substance (Road Transport Workers) (GB) Regulations 1970.

(68) The standard Transport Emergency (TREM) Card would suffice.

(69) The Dangerous Substances (Conveyance by Road in Road Tankers and Tank Containers) Regulations 1981, effective 1st January 1983.

(70) Construction and Use Regulations 1978, section 81(4).

Fig. 18 DESIGN OF 'HAZCHEM' HAZARD WARNING PANELS

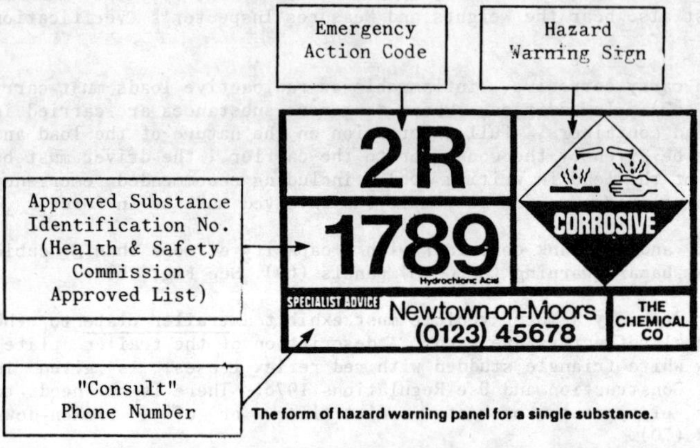

The form of hazard warning panel for a single substance.

A warning panel for a mixed consignment.

9 Plating, Testing and Type Approval

9.1 Plating and Testing

The Road Traffic Act 1972 (1) gave the Minister of Transport wide powers to require the marking of particulars on motor vehicles and trailers, and their testing and inspection. In 1964 an amendment to the then current Construction and Use Regulations introduced the concept of 'plating', the affixing by the manufacturer of a plate to vehicles and trailers. This became known as the 'Manufacturer's Plate'. The Road Safety Act 1967 (2) introduced the concept of the affixing of a plate by the Ministry of Transport. The details of this system, known as 'Ministry Plating', were laid down in the Goods Vehicles (Plating and Testing) Regulation (3).

With certain exceptions, the Plating and Testing Regulations apply to all rigid goods vehicles over 30 cwts (1525 kg) unladen weight, all drawbar trailers over 1 ton (1020 kg) unladen weight, and all articulated combinations, both drawing vehicle, semi-trailer and 'converter dollies', of any weight. The exceptions are clearly set out in Schedule 2 to the regulations. The more important of these exemptions from an operator's point of view are:- dual-purpose vehicles, motor tractors, locomotives, and vehicles so constructed that they can only be used under the Motor Vehicles (Authorisation of Special Types) Order.

9.1.1 Manufacturers' Plates and Design Weights

Although the Manufacturers' Plates were the forerunners of the Ministry Plating system, they cannot be ignored. In the first place, under the Construction and Use Regulations 1978, regulation 95, it is an offence to exceed the weight limits shown on a Manufacturer's Plate affixed to a 'Post 68' vehicle (defined below).

Under the C & U Regulations 1978 (4), Manufacturers' Plates are required to be affixed to all goods vehicles (excluding dual-purpose vehicles, tractors and locomotives) first registered on or after 1st January 1968 (referred to hereafter as 'Post 68' vehicles). They may <u>also</u> be affixed to a goods vehicle (5):-

(a) Registered before 1st January 1968;
(b) Having brakes with a 50%/25% efficiency rating (see Chapter .8, 'Braking'); and
(c) Having tyres of a sufficient ply-rating to support the weight marked on the

---

(1) Section 40.
(2) Now consolidated in the Road Traffic Act 1972, sections 45 onwards.
(3) Made in 1968, consolidated as the Goods Vehicles (Plating and Testing) Regulations 1982.
(4) Regulation 42.
(5) Regulation 85 relating to Motor Cars and Heavy Motor Cars; Regulation 86 relating to Trailers.

Manufacturer's Plate.
Such a vehicle, equipped with a Manufacturer's Plate, is referred to as a 'Prior 68' vehicle.

The Manufacturer's Plate must show (6) the manufacturer's name, engine (7) and chassis numbers, number of axles, and maximum design weights. These latter are the gross, axle and train weights within which the manufacturer considers that the vehicle is fit to be used. On 'Prior 68' vehicles there was a difference in the effect of carrying a Manufacturer's Plate and a DTp Plate; the former was advisory, while the latter was mandatory. This difference was reflected in the wording of the regulations, which spoke of Manufacturers' Plated Markings and DTp Plated Weights. Since all 'Prior 68' vehicles now have a DTp Plate the difference is academic; on 'Post 68' vehicles the question will not arise because of the effect of Regulation 95 (above), which in effect makes Manufacturers' Plates mandatory on these vehicles as well (8). In this connection, however, it is important to notice that where a Manufacturer's Plated Weight is not the same as the C & U weight limit for that vehicle, the lower of the two weights must not be exceeded.

9.1.2 DTp Plating and Testing

Motor vehicles and trailers become due for Plating and Testing in accordance with a timetable which shows
(a) The date after which the vehicle cannot legally be used without a Testing Certificate; and
(b) The last date of booking (at least one month before (a) above).

The programme for the Plating and Testing of 'Post 68' Motor Vehicles is based on an 'anniversary system', thus Motor Vehicle first registered in April of one year must be tested by the last day of April in the following year. But with Trailers the relevant factor is the date on which a trailer is sold or supplied by retail (not manufactured), and the Plating and Testing programme is then the same as for motor vehicles. Trailers, being unregistered, are allocated a number by the Goods Vehicle Centre, Swansea, and are to be submitted for First Examination in the month corresponding to the last two digits of this number (thus 03 = March). However, vehicles and trailers not registered or sold by 1st July in the year following manufacture must then be Plated and Tested by 31st December next.

Applications for test should be made on Form VTG1L (for motor vehicles) and VTG2L (for trailers). Part I of the form is the actual application for a test; Part II is for giving particulars of the vehicle and its equipment. In the case of a First Examination it is helpful, though in no way legally obligatory, if Part II is submitted earlier than the obligatory one month before the test date; this is known as pre-registration. In return the DTp will make every endeavour to allot a test booking applied for not less than 14 days before the date sought (9). There are also provisions for operators with large numbers of trailers to spread the test dates over a period, either by allocating phased serial numbers, or by pre-registering trailers and arranging for these to accompany motor vehicles to the

---

(6) Motor Vehicles (Construction & Use) Regulations 1978, schedule 2.
(7) In the case of a motor vehicle manufactured after 1st October 1972, the engine power rating must be shown. Motor Vehicles (Construction & Use) Regulations 1978 section 44 defines a minimum power-to-weight ratio as 4.4 kilowatts per 1000 kilograms GVW.
(8) This is important during the first year of a 'Post 68' vehicle's life, before its first examination.
(9) Guide for Vehicle Operators, DTp leaflet VTG20. The GVC retains the fee and passes the application to the testing station chosen, which books the test and notifies the operator.

test.

The Regulations define various types of test as follows.

A First Examination is where the vehicle is given its initial test and is plated.

A Periodic Test is in effect the vehicle's annual re-test, and will not normally involve Plating. Operators can voluntarily send motor vehicles, but not trailers, for re-test earlier than their due date; but where this is done, the next period will be on the anniversary of the issue of the last certificate.

Part II and Part III re-tests are required where the vehicle has failed a First Examination or a Periodic Test respectively.

A Part IV Test is an examination carried out after a notifiable alteration (see below).

The test itself is carried out at a DTp Testing Station. The vehicle should be accompanied by a driver who must be present throughout the test (unless permitted to be absent by the examiner) and must drive the vehicle and work the controls if required to do so. The vehicle will be inspected in accordance with criteria laid down in the MOT Goods Vehicle Tester's Manual (HMSO), the test resolving itself into four main parts viz:-
(a) Tests carried out on a hard standing;
(b) Tests carried out with a vehicle over a pit;
(c) Lighting check;
(d) Brake testing.

If the vehicle fails the test, the driver will be handed a Notification of Refusal of a Goods Vehicle Test Certificate, Form VTG4, detailing the defects for which the vehicle has been failed. If the vehicle is presented for retest by arrangement (10) the same or following day, no fee will be payable. If the vehicle is presented with fourteen days (11), a reduced fee will be payable. After that time, or if the vehicle is presented at another Testing Station, a fresh application must be made.

9.1.3 Preparation for the DTp Test

Ensure that the vehicle is clean and that an appointment has been made for the vehicle to be tested. The relevant documents should accompany the vehicle.

Preliminary inspection If regular maintenance, based on the DTp procedure, has been followed, then this inspection will be confirmation of work done rather than an inspection prior to an extensive repair schedule. However, it is as well that this inspection is done about a month prior to the actual test; thus faults which are brought to light may be rectified in good time.

Broadly, the inspection may be divided into the following main groups.

Bodywork should be sound and securely fastened to the chassis. Check securing bolts with a crowbar. Ensure that no sharp projections exist and that the driving

---

(10) This is only permitted where the failure is due to certain prescribed minor defects.
(11) The retest can be applied for to, and held at, a station of the operator's choice up to 7 days after the failure. Applications after 7 days but within 14 days must be made to the station where the vehicle was submitted for the previous test. (Goods Vehicles (Plating and Testing) (Amendment) Regulation, 1105, 1973).

seat is secure.

Chassis frame - check that this is free from corrosion and cracks.

Suspension - check all anchor brackets, bolts, springs etc.

Steering - check for overall play in the system, and also for wear in the kingpins. Free use should be made of the crowbar in order to determine where wear exists. Jack wheels off the ground to check play in bearings and joints.

Lights - visually check the wiring system. Switch on each circuit in turn and check that all lights are working.

(Note - send spare bulbs with the vehicle. The driver is permitted to rectify minor faults while the vehicle is being inspected).

Headlights may be aligned with a wall diagram. Where a trailer is involved, ensure that couplings are in good order.

Brakes should be visually inspected for wear and/or other defects. They should be correctly adjusted.

With the air system empty, remove the drain tap of each tank and replace with a pressure gauge. Bring the pressure in the system up to normal operating pressure and stop the engine. Apply the brake pedal and hold on. Check that there is no pressure loss in the system. Leaks may be located by ear or, if in doubt, then a soap solution should be applied at possible leakage points. Check the efficient operation of Main, Secondary and Parking Brakes under full load conditions - preferably with the use of a brake meter.

Check that warning lights and/or buzzers are operating satisfactorily.

The foregoing is a resume of the main points of an inspection, but the various trade organisations issue or make available excellent inspection reports. Regardless of the fact that a trade report, or a private inspection service, is used, it is essential for the user to ensure that the inspection was done **METHODICALLY**, and that items checked are ticked as they are done. All items checked should be compared with the DTp list.

9.1.4 Notifiable Alterations

Any alteration to the structure or equipment of a vehicle which might change the vehicle's carrying capacity, making it unsafe to operate at the plated weight, or which might adversely affect the braking performance, must be reported to the GVC at Swansea, and the DTp may then decide that the vehicle should be submitted for re-examination for plating. Alternatively the operator may request re-examination with a view to obtaining a higher plated weight. In the latter case a fee is payable.

9.1.5 Tyres

An operator should ensure that the tyres with which a vehicle is fitted are capable of supporting the full plated weight of that vehicle (12). In certain cases, however, an operator, because of the nature of the work he does or the loads he carries, may wish to fit tyres of a lower ply rating than that shown in the Table (13) giving the ply rating, size and pressure necessary at different

(12) Motor Vehicles (Construction and Use) Regulations 1978, sections 148/149.
(13) Supplementary Type Loading Table GP2 (HMSO) or Table 2B of SMMT Tyre and Wheel Alignment Manual.

plated axle weights. If he does so and notifies the fact to the examiner, he will be given a DTp Plate for a lower plated weight, and the plate will be endorsed either GP/1 (Speed not to exceed 40 m.p.h.) or 2J (Speed not to exceed 40 m.p.h. and use of vehicle to be confined to 25 miles radius of the operating centre).

9.1.6 Appeals

Appeals against refusal of a test certificate, refusal to carry out a test, or against the weight at which a vehicle has been plated, can be made initially to an Area Mechanical Engineer of the DTp (within 10 days on Form VTG8); or finally appeals against the Engineer's decision can be made to the Secretary of State himself (within 14 days on Form VTG9). There are prescribed fees of £15 and £25 respectively, refunded if the appeal is sustained, or if the Secretary of State considers that there were substantial grounds for appeal.

9.1.7 Points to Watch when Reporting for Test

(a) Take the examination appointment card and check date and time.
(b) Ensure the vehicle has sufficient fuel.
(c) See that the driver carries any documents asked for on the appointment card, e.g. registration book (motor vehicle) or evidence of date of manufacture (unregistered vehicle).
(d) Ensure that any identifying mark, e.g. trailer number, is clearly stamped.
(e) If possible, send someone with mechanical knowledge who may be able to carry out simple repairs, e.g. fitting a new light bulb, at the time of test.

9.1.8 Reasons for Refusal to Conduct a Test

The DTp may refuse to conduct a test for any of the following reasons (14):-
(a) The vehicle is late for the test.
(b) The examination appointment card and any documents required to accompany it are not produced. The Registration Document (Form V5) or other proof of vehicle identity is required.
(c) There is a discrepancy between the vehicle presented, and the particulars given on the application.
(d) The vehicle is not presented with a trailer if booked with one; or the vehicle is presented without a load if this is requested on the appointment card.
(e) Identification marks are not clear, or non-existent.
(f) The vehicle is unreasonably dirty.
(g) There is not enough fuel to complete the test.
(h) The vehicle is not laden if this is requested; or is not made safe (and so certified) if it normally carries a dangerous load.

9.1.9 Criteria used in assessing Vehicle Plated Weights

Schedules 6 and 7 of the C & U Regulations 1978 lay down maximum gross vehicle weights for particular types of vehicles and combinations of vehicles. A vehicle will not be plated for a weight in excess of these limits, even if it carries a manufacturer's plate showing a higher design weight.

Standard List This is a list, prepared by the DTp and obtainable from HMSO, showing in respect of goods vehicles by make, type and model the design weights and the weights at which the vehicles will be plated. The list is prepared in consultation with interested parties, and is sectionalised so that a particular maker's list can be bought. (The maker may publish a copy of the list in respect of his own vehicle, and this will be a useful guide; but in any case of dispute

---

(14) Goods Vehicle (Plating and Testing) Regulations 1982.

the standard list published by HMSO will be referred to).

Manufacturer's Plate This will invariably show the same design weight as the standard list.

First Examination In the event of the vehicle not being of a make, type or model listed in the Standard List, the vehicle examiner may determine its plated weight, having regard to any information available regarding the actual vehicle (i.e. as supplied by the operator on Part II of forms VTG1L and VTG2L) and the weight at which it was originally designed to operate, any alterations, the braking effort achieved, and certain Construction and Use criteria.

The DTp Plate (VTG6) will show the Plated Gross, Axle and (in certain circumstances) Train Weight of the vehicle, and must be affixed either in the cab or, in the case of the trailer, in a suitable position (preferably on the nearside) (15).

9.2 Goods Vehicle Type Approval

All Goods Vehicles (but not trailers)
(a) manufactured on or after 1st October 1982, and
(b) first used on or after 1st April 1983,
must now be 'GB Type Approved' by the manufacturer (or importer). (Type Approval is not the responsibility of the operator).

The purpose of Goods Vehicle Type Approval is to:-
(a) check that new vehicles comply with specific standards listed in Schedule 1 of the Motor Vehicles (Type Approval for Goods Vehicles) (GB) Regulations 1981. These are EEC, ECE and C&U standards. This individual systems approval is an essential prerequisite of Type Approval;
(b) determine, where appropriate, maximum gross train and axle weights.

Goods Vehicle National Type Approval applies to:-
(a) all goods vehicles (but not trailers);
(b) motor caravans and motor ambulances:
(c) bi-purpose vehicles, i.e. those constructed for carriage of both goods and eight or less passengers, and not included in passenger car Type Approval.

Vehicles 'exempt from' Goods Vehicle Type Approval include:-
(a) Trailers;
(b) Vehicles licensed abroad which are temporary imports;
(c) Vehicles travelling for export;
(d) Vehicles in the service of visiting forces;
(e) Vehicles which are, or were formerly, in use in the public service of the Crown;
(f) Prototype vehicles;
(g) Motor tractors, light and heavy locomotives as defined in the Road Traffic Act 1972;
(h) Engineering plant including:-
    (i) pedestrian-controlled vehicles;
    (ii) straddle carriers;
    (iii) works trucks and track-laying vehicles as defined in the Construction and Use Regulations;
(i) Special type vehicles as defined in the Motor Vehicles (Authorisation of Special Types) General Order 1979;
(j) Tower wagons as defined in the Vehicle (Excise) Act 1971;
(k) Fire engines;

---

(15) Motor Vehicles (Construction & Use) Regulations 1978, sections 148/149.

(l) Road rollers;
(m) Steam propelled vehicles;
(n) Gritting and snow-clearance vehicles;
(o) Electrically propelled vehicles;
(p) Breakdown vehicles;
(q) Vehicles not exceeding 1515 kg unladen weight, and not part of an artic combination, which are either 'personal imports' or personally assembled.

9.2.1 Component Approval

In order to fulfil the purpose of Goods Vehicle Type Approval, quite complex construction standards have to be met by the manufacturer. These standards are referred to in a table embodied in Regulation 5 of the Construction and Use Regulations 1978, and are either
(a) An EEC (European Community) standard; or
(b) An ECE (European Council of Ministers of Transport) standard; or
(c) A Construction and Use standard.

The above standards are component or systems standards; they do not cover complete vehicles. Where the UK has accepted an ECE standard, or adopted (under the EEC Type Approval Directive) an EEC standard by incorporating it into its own legislation, then a vehicle Type Approved by reference to that standard may be used in the UK and, in the case of standards (a) and (b) above, will be deemed to be exempt from the equivalent Construction and Use (c) standard.

Component Approval, either
(i) GB Component Approval (where no International standards (a) or (b) exist); or
(ii) Full International Component Approval,
is obtained by the manufacturer/importer from the DTp's Vehicle and Component Approval Division at Bristol, or the Vehicle Type Approval Centre attached to the Motor Industry Research Association (MIRA) at Nuneaton. Components passing this standard are marked as follows:-

E11 — ECE Approval Mark

e11 — EEC Approval Mark

Once approval of a system/component is given, it becomes illegal to fit parts not having an approval mark to such systems on a vehicle.

9.2.2 Vehicle Approval

Once a manufacturer has met Component Approval standards in respect of all the systems on his vehicle which are required by the Goods Vehicle National Type Approval to have approval, he may apply to the Vehicle and Component Approval Division (VCA) at Bristol for Goods Vehicle National Type Approval.

In consultation with VCA, the whole list of possible variations of the model is considered and a 'worst case vehicle', incorporating all those features which would most adversely affect Type Approval standards, is submitted. If it passes, then any vehicle built to any combination of the variations can be Type Approved. There are two types of Type Approval:-
(a) Type Approval Certificate: for EEC vehicles, issued in respect of a vehicle type. The manufacturer then issues the dealer or operator with a Certificate of Conformity (except for rigid vehicles under 1525 kg, covered by Motor Car Type Approval).

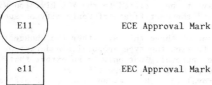

(b) Minister's Approval Certificate for any vehicle, including non-EEC vehicles. It is issued in respect of a specific vehicle. The DTp can issue a subsequent MAC to other identical vehicles.

9.2.3 Registration, Licensing and Plating

The Certificate of Conformity or subsequent MAC issued by the manufacturer/importer to the dealer/buyer is needed for first registration and plating. The flow chart in Fig. 19 illustrates how the paperwork to achieve this circulates. Note how the third copy of the Certificate of Conformity/subsequent MAC acts as a temporary plate until GVC Swansea issues the DTp plate (well before the anniversary of the vehicle's first registration).

9.2.4 Alterations to Type Approved Vehicles

Where a manufacturer proposes to make an alteration to an approved basic specification (the so-called Type 1 alteration to a 'frozen' specification), he must either apply to VCA to vary the TAC, or to issue a new TAC, depending on whether the alteration creates a new variant outside the 'worst case'.

Where a Certificate of Conformity or subsequent MAC has already been issued but the vehicle has not yet been registered, the procedure to be followed depends upon whether this so-called Type 2 (post certification/pre registration) alteration comes within the definition in the Type Approval Regulations of a Prescribed Alteration. At present, Prescribed Alterations are only those which:-
(a) Change the number of axles;
(b) Change the position of the axles (wheelbase); or
(c) Change significantly wheel dimensions, e.g. fitting smaller wheels or low-profile radials to reduce height, or super single tyres in place of twin wheels.

If any Prescribed Alterations are made to a type approved vehicle before it is first registered, they have to be notified to the VCA Division with appropriate form and fee to be paid, and the Certificate of Conformity or subsequent MAC.

The VCA will then decide one of three things. Either the changes:-
(a) have no significant effect on the type approval standards; or
(b) they do affect type approval, but only to an extent that the particular system or systems altered will need inspecting; or
(c) they affect type approval to such an extent that the whole vehicle needs to be tested again. (See Fig. 20).

In the first case, the VCA will alter the Certificate of Conformity or subsequent MAC, send it back, send a copy to the Goods Vehicle Centre at Swansea, and the owner can go ahead and license the vehicle. In the second case, the VCA will suspend the certificate and the person carrying out the alterations has to apply (with another fee) for approval of the altered systems.

In the last case, the certificate is cancelled, and the person carrying out the alterations would have to apply for new complete type approval testing of the whole vehicle.

Alterations which are not Prescribed Alterations may, under the Plating and Testing Regulations which we considered in section 9.1, be notifiable alterations if they affect the carrying capacity of a vehicle subject to those regulations. Refer back to section 9.1 for the procedure for dealing with notifiable alterations via GVC Swansea.

Where a dealer or operator makes an alteration before registering a vehicle, and is unsure how to deal with this, he should notify GVC Swansea. Manufacturers or

Fig. 19 FORMS USED IN REGISTRATION, LICENSING AND PLATING

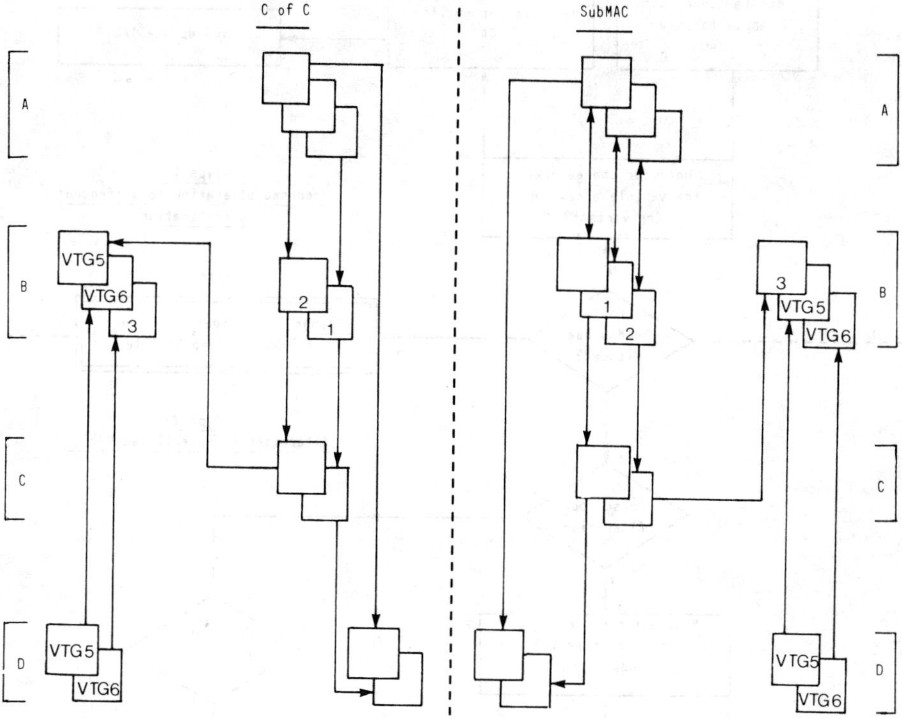

C of C

A. Original to VCA Bristol for authentification

B. Manufacturer prepares:
   1. Declaration if exempt from Plating & Testing
   2. Statement of whether any alterations made
   3. Stamped as temporary plate

C. Local Vehicle Licensing Office issues Vehicle Excise Document

D. GVC Swansea compares these documents to each other and, if satisfactory, issue <u>**Plating Certificate**</u> (VTG5) and <u>**Plate**</u> (VTG6)

SubMAC

A. Original to VCA Bristol for authentification

B. Dealer or Applicant prepares:
   1. Declaration if exempt from Plating & Testing
   2. Statement of whether any alterations made
   3. Stamped as temporary plate

C. Local Vehicle Licensing Office issues Vehicle Excise Document

D. GVC Swansea compares these documents to each other and, if satisfactory, issue <u>**Plating Certificate**</u> (VTG5) and <u>**Plate**</u> (VTG6)

Fig 20 ALTERATIONS TO TYPE APPROVED VEHICLES

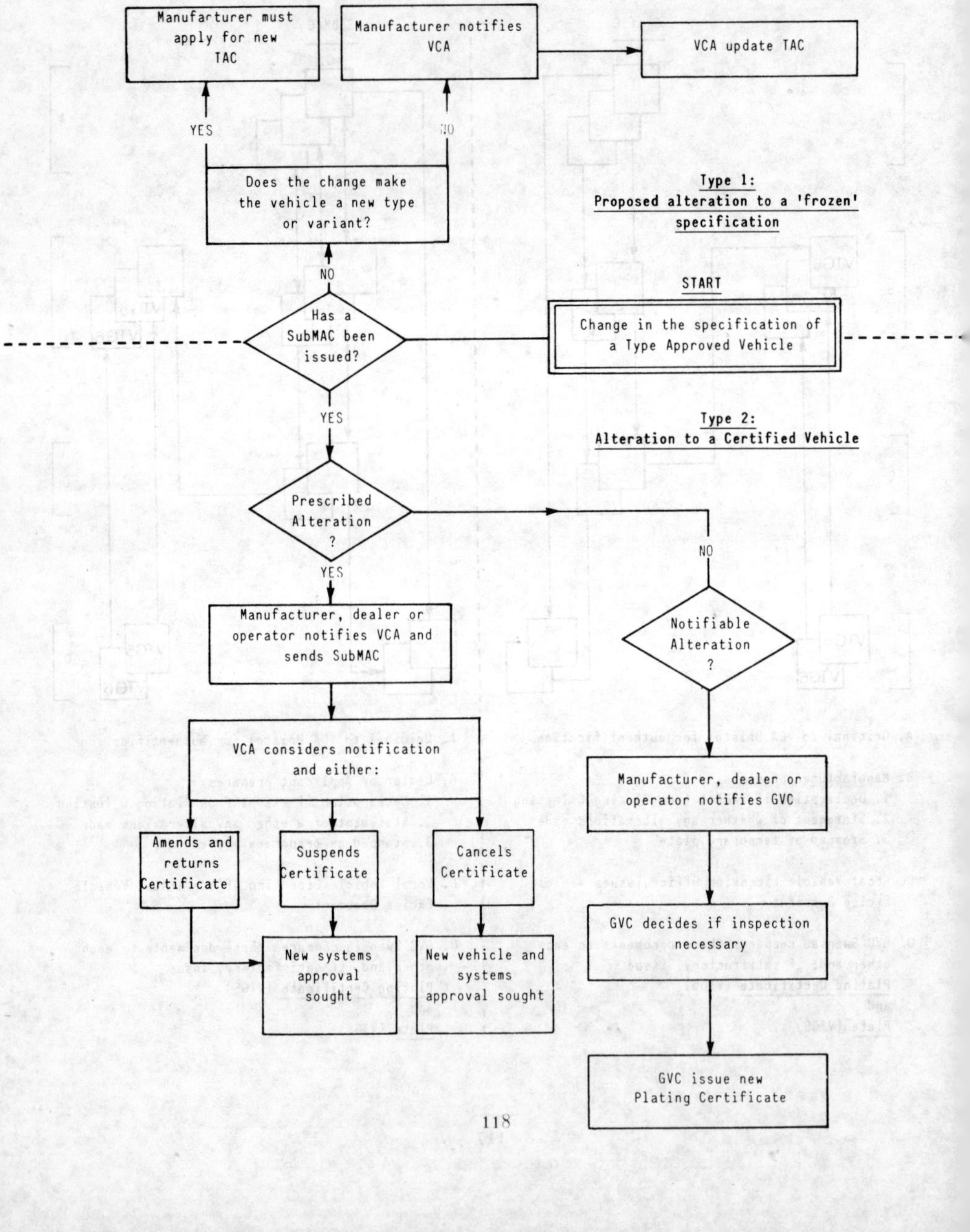

Agents similarly should notify VCA Bristol. VCA and GVC will re-route notifications to the appropriate address.

9.2.5 Alteration after Registration

Once a vehicle has been licensed, Goods Vehicle National Type Approval no longer applies. If a vehicle is subject to Plating and Testing, notifiable alterations must be advised to GVC Swansea.

9.2.6 Convertors and Body Builders

Manufacturers are not required to issue Certificates of Conformity for vehicles under 1525 kg, i.e. outside Plating and Testing Regulations. However, conversion may cause this weight to be exceeded. In this case, and in cases where a Certificate of Conformity is required, or a subsequent MAC is raised, the convertor must ensure that the dealer/operator who is to license the vehicle has the required two copies, if necessary, asking the manufacturer to raise a duplicate. Convertors must notify alterations including components/systems and notifiable alterations as above. Convertors can obtain their own Type Approval Certificate for a type vehicle incorporating their conversion by negotiating a satisfactory Conformity of Production (CoP) agreement with the manufacturer and/or his agent, and with the DTp.

10 Fleet Management and Costing

10.1 Business and Financial Management

Business management could be described as the task of directing all the human affairs (1) which concern the running of a business. In the case of a road goods transport operator, the 'business' is the provision of a service which is traded for revenue.

10.1.1 Fundamentals of Management

The effective Manager performs a number of functions, including:-
(a) **Controlling** on-going activities;
(b) **Organising** the application of the resources under his control, including Labour; Finance; Vehicles and equipment;
(c) **Planning** future activities;
(d) **Co-ordinating**, directing and monitoring the activities under his control, so as to achieve his objectives in a way which utilises his resources most efficiently (2).

Managers will deploy certain techniques to help them. Amongst these are:-

(1) Delegation of authority (not responsibility) in order to free the Manager from routine tasks and allow him to concentrate on his 'key results' areas (3). The more senior the Manager, the more he will tend to delegate functions (a) and (b) above, in order to concentrate on functions (c) and (d).

(2) Motivation of staff in order to maximise their potential. This means creating good working conditions (4), good industrial relations, and a climate of involvement (5).

(3) Discipline. An obviously fair disciplinary procedure impartially applied (6) is a pre-requisite of good road transport management because
(a) the majority of staff work on the road away from direct supervision and control; and
(b) it is essential to ensure compliance with legal obligations (7).

---

(1) P Drucker.
(2) 'Management by Objectives' - J Humble.
(3) Pareto's Law - '20% of a Manager's tasks produce 80% of his results'.
(4) Hertzerg's Heigiene Theory (q.v.)
(5) Maslow's Hierarchy of Needs (q.v.)
(6) Code of Industrial Relations Practice, HMSO.
(7) Road Traffic Act 1972, in particular section 40 (Construction and Use of Vehicles), section 160 (Weighing of Vehicles) and part III (Driver Licensing); Transport Act 1968, section 96 (Drivers' Hours and Records).

(4) Communication. Again because of the lack of direct supervision within the industry, good communications in a business is essential. How this is done depends very much on the circumstances, but as a minimum, management should ensure that

(a) information is imparted as clearly and unambiguously as possible;
(b) there is no obvious factor likely to interfere with the reception of the information (e.g. memos not copied to all concerned); and
(c) there is adequate 'feedback', i.e. there is opportunity for the information to be questioned or commented upon.

Note that while the elected representatives of the employees may usefully assist in providing 'feedback' to a management communication, it is not their job, nor what they were elected to do, to communicate management's decisions to their constituents - even if they were party in negotiation to the decisions (8).

10.1.2 Profit and Loss Account and Balance Sheet

The Profit and Loss Account of an operator will record the success or otherwise of his operations during a trading period. All revenues will be credited and all expenses will be debited. The deduction of direct trading costs (wages, fuel etc.) will produce a Gross Profit (9) from which overhead expenses are deducted to arrive at a Net Profit. This is then appropriated, either by distribution as drawings or dividends, or adding to the business's assets (either as a reserve or by writing down (10)), or by conversion to a fixed asset (e.g. a vehicle).

The Balance Sheet, drawn up at the end of a trading period, is a statement reflecting the business's financial position at that date. It shows the liabilities, i.e. what is owed by the owners to investors and lenders including, in the case of proprietors and partners, themselves (as capital invested, dividends etc), to the creditors (e.g. the suppliers of fuel), and to the Inland Revenue (Corporation Tax); and the assets, either fixed (depots, vehicles, stock etc. at present book value) or current, i.e. easily convertible to cash (debtors - who owe the business money - bank deposits, cash and stock in hand).

10.1.3 Capital

The capital employed in a business is represented by the total assets of the business less its current liabilities. It represents the owner's financial stake in the undertaking, i.e. the business's long-term funding - including any share capital, reserves and long-term loans. The ways in which the capital is raised will depend on

(a) the type of ownership (11);
(b) the long-term or short-term use to which the capital is to be put.

Among the more common sources of capital are

(1) Proprietor's and partners' capital invested;
(2) Hire-purchase of an asset (e.g. a vehicle);
(3) Bank loans and overdrafts;
(4) Debentures issued against the security of the assets in general, or of a particular asset;
(5) Share capital (12).

Note that the necessity of increasing the capital employed in a business may be

(8) A Forest (Industrial Society), 1969.
(9) Often drawn up as a separate 'Trading Account'.
(10) e.g. by repayment of a loan (liability).
(11) See Chapter 11 (e.g. a sole proprietor cannot raise share capital).
(12) Subscribed within the limitations of the company's authorised capital.

avoided by leasing (13) or hiring (14) the required asset (e.g. vehicles).

Banks provide a number of useful services to the operator. As well as conducting the business's account by receiving payments in of revenue and honouring cheques (15) drawn on the account by the operator, the bank will agree the terms on which the account may be overdrawn (16) in order to overcome a temporary shortage of working capital (17). Banks are also of course lenders of finance, either by extending overdraft facilities as above or by arranging commercial loans. In either case they naturally charge interest.

<u>Working Capital</u> is represented by those assets which are easily realisable (essentially cash, bank credit balances, and debtors). It is of considerable importance to the operator since it should be at least equivalent to his current liabilities to avoid insolvency. It can be calculated by deducting current liabilities from current assets.

10.1.4 Cash Flow

This is the movement of cash into and out of an enterprise. Net cash flow is the difference between cash received and cash paid, and where this becomes negative the enterprise becomes insolvent - even though there may be no question of bankruptcy (i.e. total assets are not less than total liabilities).

Solvency is as important as profitability if the undertaking is to survive, and one way in which to guard against harmful negative cash flows is to prepare a budget indicating the anticipated size and timing of all income and expenditure, based on expected levels of trading activity.

<u>Budgetary Control</u> is of further use in that it enables an operator to spot variances from the agreed budget, seek an explanation for these, and if necessary to take corrective action, including updating the budget (18).

It will often happen that a present cash outflow (e.g. the purchase of a vehicle) is intended to produce a future benefit in the form of revenue inflow, justifying the investment over the life of the asset (19).

10.1.5 Management Ratios

Managers traditionally compare certain figures in operators' Balance Sheets in order to gauge both their profitability and their liquidity.

---

(13) The lessee does not own the vehicle during the primary period of the lease (after which there is often, as with HP, an option to purchase), but is responsible for all operating costs.

(14) The hire company usually undertakes repairs, maintenance and supply of replacements.

(15) A cheque is technically a Bill of Exchange, whereby the drawer instructs his bank to immediately pay to the credit of the drawee the amount promised.

(16) Known as an overdraft.

(17) Technically insolvency.

(18) This technique is a particular example of 'Management by objectives', and is known as 'Management by exception' or 'Exception reporting'.

(19) Since a future net inflow is, at current values, worth less than the same inflow today, the cash flow is <u>discounted</u>, e.g. £100 invested today at 10% will be worth £110 in one year's time. Thus a net inflow of £110 in one year's time is discounted as £100 at today's prices for comparative investment appraisal purposes. This technique is called Discounted Cash Flow (DCF).

(a) **The Liquidity (Acid Test) Ratio** is calculated as $\frac{\text{Liquid Assets}}{\text{Current Liabilities}}$.

Liquid Assets include only assets quickly realisable - essentially debtors (20) and cash in hand and at bank. Current liabilities are such items as tax due, dividends payable, and creditors. None of these are deferrable, and if the working capital is insufficient to cover them the operator will be insolvent. The recommended ratio is above 1.

(b) **The Working Capital or Current Ratio** is calculated as $\frac{\text{Current Assets}}{\text{Current Liabilities}}$.

Current assets include working capital and those other assets such as stock which could in theory be realised for cash. The current ratio is not so good an acid test as the liquidity ratio, and the recommended ratio is about 2.

(c) **Return on Capital Employed** is calculated as $\frac{\text{Net Profit}}{\text{Capital Employed}} \times 100\%$.

This is an obvious measure of profitability. If the Halifax Building Society were currently paying 7.5% interest after tax, and if the operator's return on capital employed was 5%, he would be better off realising his capital and investing in the Halifax - assuming that his objectives were purely financial, which is by no means always the case.

10.2 Commercial Aspects

The commercial objectives of an undertaking will depend upon its type of ownership and control, and the 'return' expected on the capital employed will be measured by various yardsticks, including such diverse objectives as profit maximisation, least cost public service, retention of employment etc.

10.2.1 Revenue

The payments made by customers for the services with which they are provided is understandably of vital importance to the operator. Prompt invoicing and good credit control can do much to avoid unnecessary liquidity problems.

The management ratio (21):- $\frac{\text{Debtors (22)}}{\text{Sales (23)}} \times 12$

gives the average time (in months) which it takes customers to settle their accounts. Revenue is applied to service:-

(a) Capital (as dividends or reserves for investment);
(b) Labour (as remuneration);
(c) Operations (by purchase of materials - fuel, oil, tyres, stock, vehicles etc.)

10.2.2 Documentation

It is necessary in any business to accurately record the daily activities and transactions. Much of the information may be common to more than one document, and the operator should therefore consider

(a) 'Aligning' his documents in a common format so that the same information appears in the same place on every form (e.g. consignee's name and address

---

(20) excluding bad debts and any cash discounts offered.
(21) Debtors'. Ratio.
(22) From the Balance Sheet.
(23) Sales, Turnover or Revenue from Trading Account and Profit & Loss Account.

always in top left hand corner);

(b) the use of Electronic Data Processing to retain and retrieve 'banks' of data such as customer's accounts, sales ledgers, payroll, deliveries etc.

The following documents will feature in some form in an operator's system.

Quotation - an offer (which may be verbal or in writing) of a price for a specified service.

Order - an acceptance of a quotation received from a customer, or initiated by the operator (e.g. a fuel order), or by a customer for the operator's services.

Invoice - a request for payment for a service rendered.

Credit Note - a notification of a deduction from the customer's account balance in respect of some alteration in the transaction, e.g. goods returned, claims etc. Similarly a Debit Note may be raised to rectify an undercharge (24).

Statement - a notification to a customer of the state of his account with the operator. In effect, a copy of the customer's entry in the sales ledger, showing invoices debited and payments credited.

Consignment Note - a document describing a delivery and showing the name and address of the consignor and consignee, and details of the consignment. If every consignment is entered on a separate note this is known as the single entry system, whereas the multiple entry system involves entering all the consignments in one load on to one note; customers sign for proof of delivery purposes on the same line as the entry describing their delivery. Consignment notes are often prepared in sets with copies for carrier (to file), consignee (to retain with goods) and driver (to obtain signature). They have a legal significance under CMR (25) and EEC (26) legislation; they must be carried with those loads to which the Act and Regulations apply.

10.2.3 Sub-Contracting

It is obviously not sensible for an operator carrying for hire and reward to run a fleet sufficiently large to satisfy all his customers' demands all of the time. There will be peaks of demand, and these will dictate various strategies.

Sub-contracting is the practice of accepting an order and then passing this to another contractor, because the main contractor does not have sufficient capacity. Usually the sub-contractor is paid 90% of the rate, the main contractor retaining 10% commission. The confirmation note from the main contractor to the sub-contractor generally provides that the latter will carry under the conditions of carriage agreed with the consignor (27).

Clearing Houses, as the name implies, are a type of haulage broker who will provide a client's load for a carrier who is looking for work (possibly a 'back load'). The client could (as above) be an operator trying to sub-contract additional work (28).

Freight Forwarders are also able to offer a service to an operator accepting a

(24) e.g. a Demurrage charge may arise if a vehicle is unreasonably detained.
(25) Carriage of Goods by Road Act 1965.
(26) EEC 'Forked Tariff' Regulations 358/69.
(27) RHA Conditions of Carriage, 1967 and 1982.
(28) Commission of 10% is incorporated in the standard Clearing Houses Conference agreement.

load to or from overseas, by arranging customs clearances, documentation, overseas agencies, packing etc.

Groupage Freight forwarders and some hauliers offer groupage facilities to consignors having less than container-loads (or less than vehicle loads) for a destination. These are consolidated with other consignors' loads to make up full loads.

Hired Transport may be resorted to by the operator on either a long-term or a casual basis. If the operator is an own-account operator (29), he may 'spot hire' haulage to make deliveries at peak times over and above the capacity of his own fleet, or he may contract hire vehicles with or without drivers (30).

The advantages of hiring are:-

(a) The operator's capital is not tied up in the asset (vehicle) and can be put to work in other profitable areas of the business (e.g. warehousing);
(b) The operator is relieved of the responsibilities of maintaining the vehicles (31);
(c) The hiring company will frequently agree to replace a broken-down vehicle;
(d) The operator's costs are known in advance, since they are based on the terms of the hire contract, and are not affected by unpredictable items like sudden unexpected repair bills.

The disadvantages of hiring are:-

(a) The hirer does not always have the same degree of control over hired vehicles (and in some cases drivers) as he does over his own fleet;
(b) Unless the hire company agrees to paint the vehicles in the hirer's own livery, the advertising effectiveness of the fleet is not utilised;
(c) The total cost of hiring may exceed the cost of owning, since the hire company will require a return on their capital employed.

10.3 Costing

Cost accounting is the recording of actual costs as they occur, and allocating them to causes, i.e. to resources, as they are employed. The financial accounts of an operator are necessarily historic, and of little use for monitoring and controlling his operations.

10.3.1 The Purpose of Costing Systems

A costing system is not a panacea, but it can only point the way for the Manager - in particular:-

(a) It is essential as a guide to rates.
(b) If the operator does not know costs he cannot know profit.
(c) A costing system is essential to a Manager in a fleet replacement programme.
(d) An operator needs to know not only what his costs are, but also whether or not they are at the correct level.
(e) A good costing system will reveal poor operation, inefficient vehicles or mis-management, so that corrective action can be taken in time.

---

(29) i.e. carrying only goods in connection with his trade or business. Note that many own-account operators will require a standard O-licence because they may wish to back-load or to carry for a subsidiary company.

(30) Note that the employer of the driver must hold the O-licence on which the vehicle is specified.

(31) If he holds the O-licence, he is still responsible for ensuring that the vehicle is regularly inspected and that records are kept of inspection; he is also responsible for its condition on the road.

The costing system used must fulfil the following functions:-
(a) To break down costs into standing costs, running costs, and wages.
(b) To show such costs for each individual vehicle.
(c) To compare costs with revenue to show profitability.
(d) To be capable of showing up changes in costs and/or profits.
(e) To be capable of showing vehicles or drivers which are not being efficiently utilised.

10.3.2 Methods of Costing

The essential feature of a good costing system is that it should record all costs incurred, so that these may be recovered by incorporation into the rates which the operator quotes. How the individual costs are actually classified will be a matter for each operator, depending on his operating methods, but a suggested classification of vehicle costs is as follows.

Standing Charge (incurred whether the vehicle is operating or not)

(a) Direct Charges (i) Depreciation on vehicle and its share of trailers etc.
(ii) Road Fund Tax (Excise Licence)
(iii) O-Licence Fee
(iv) Insurance of the vehicle and load.

(b) Establishment Charges

Salaries	Telephones	Other Insurance
Holiday Pay	Stationery	Entertaining and Travelling
Rent and Rates	Postage	Advertising and Publicity
Training Levy	Subscriptions	Legal and Professional Fees
Maintenance Overheads	Office Cleaning	Interest on Capital Employed
Bad Debts	Light and Heat	

The above are totalled and divided by the total carrying capacity of the fleet to obtain annual charges per ton. (Multiply this figure by the carrying capacity of each vehicle to obtain that vehicle's share of the charges).

Having calculated an annual standing charge, this must now be converted to a daily charge. This is done by calculating the number of working days per year (on a 5-day week basis allowing for Public Holidays though not for annual holidays), and dividing the total annual standing cost by this figure.

Running Costs (directly related to mileage)
(i) Fuel
(ii) Oil
(iii) Maintenance
(iv) Tyres

Wages and Expenses (including Employer's Contributions to National Insurance)

The amount to be included in the standing costs for depreciation can be found by reference to the Profit and Loss Account. Usually the cost of putting the vehicle on the road (32) less its estimated residual (scrap) value is spread over the vehicle's anticipated life either
(a) on the basis of a fixed amount per year; or
(b) by debiting each year a percentage of the written-down value (33).
Neither (a) nor (b) above take account of the effect of inflation on vehicle

---

(32) Less its tyre costs, which are recovered through the running costs.
(33) This 'Reducing Balance' method is often preferred, since it is argued that depreciation lessens as maintenance costs increase with the vehicle's age; thus the two costs in total remain fairly constant.

replacement costs, and more sophisticated methods using 'current cost accounting' principles are nowadays increasingly employed.

10.3.3 Traffic Rates

The rate for any job can be based on a costing of the time and mileage involved plus an element of profit. The time involved can be rated from a study of the relevant daily standing charge and driver's wage plus subsistence costs as incurred.

The mileage cost should also be available from the costing system employed.

A cost sheet similar to Fig.21 would be suitable for weekly or monthly use. It is recommended for weekly use, backed up by a monthly system showing actual (as opposed to estimated) costs as incurred. A monthly cost sheet would follow the same layout, but running costs would be divided into separate sub-headings to show maintenance figures, fuel figures etc. separately.

It is usual to quote rates on a round-trip basis, unless a return load can be assured.

The return on capital employed, which is built into the standing charge, must directly affect the rate quoted. While in the long run it will not be sensible to accept traffic the revenue from which does not cover all standing costs (including an expected minimum return on capital employed), the rate actually quoted will be influenced by two further factors:-

(a) The rate which the traffic will bear. Some low-grade traffics will not be offered if the full economic rate is asked, since the addition of a high transit cost to their low intrinsic cost will make them unsaleable. The operator must then decide whether to lower his rates to attract this traffic (possibly cross-subsidising it with a 'plus rate' on high-grade traffics such as perishables and high value cargo).
(b) Marginal costing. The operator with idle vehicles and drivers may in the short run offer his services at a rate which covers their running (mileage) and time (wages) costs, and makes a contribution towards standing costs.

Contract Quotations Any such quotation must be based on the weekly standby charge of a vehicle (including, if desired, the driver's wage) plus a mileage charge. Both figures should include an element of profit. A minimum mileage charge may be built into the weekly standby charge quoted, in order to guarantee the minimum return on capital.

10.3.4 Purchasing and Stock Control

Sound purchasing and stock control, particularly of fuel and spares, can enable an operator to significantly reduce his costs (and hence his rates) since this will

(a) release working capital, and decrease the interest payable on capital tied up in these assets; and
(b) release storage space which can often be utilised for more profitable activities (e.g. an ancillary warehousing facility, designed to attract additional traffic).

It is possible to work out optimum economic order quantities (34) which will theoretically help to achieve the above objectives, whilst guarding against the risks of running out of stock.

---

(34) The computation of 'Economic Order Quantities' is an Operational Research technique. It is based on the assumption that optimisation results when acquisition costs and holding costs are balanced.

FIGURE 21. Example of Weekly/Monthly Costing Sheet.

'ROAD HAULAGE LTD' Weekly/Monthly Costing Sheet Week/Month Number....

1	Vehicle Number	1	2	3	4	5	Total	Per Mile
2	Miles							
3	Standing Charge	(As calculated every 6 months)						
4	Running Cost							
5	Wages and Expenses							
6	Total Cost	(Total of Columns 3, 4 and 5)						
7	Revenue							
8	Profit or Loss	(Column 7 minus Column 6)						
9	Carrying Capacity (Tons)							
10	Repair Days							
11	Days when no work available							
12	Days when no driver available							
CUMULATIVE 13	Profit or Loss							
14	Repair Days							
15	Days on which no work							
16	Days on which no driver							
17	Mileage							

11 Law of Business and Carriage

11.1 General Principles of Law

The **Law** as it exists today is a continually developing and evolving set of rules enforceable by the State, imposed upon citizens and derived from numerous **sources**, the two most important of which to the haulier are:-

11.1.1 Common Law and Statute Law

Common Law The courts today, especially where there are no specific statutes to guide them, will rely on the decisions or **precedents** recorded in earlier cases. The Common Law comprises such a body of **case law** and is derived from many sources. For example, in the Middle Ages law was administered by travelling Assize Court Judges who discussed their cases with each other, and introduced the benefit of a degree of standardisation into recorded decisions.

A separate development was the rule of **equity** with its 'fair' remedies such as **injunctions** (restraining a defendant) and **specific performance** (directing a defendant to do his duty). It was channelled into the Common Law by the Judicature Acts 1873 and 1875, which merged the separate Court of Chancery (which granted remedies under equity) into a Division of the High Court. The Acts also provided that 'in any dispute between the Common Law and Equity, Equity prevails'. Thus today equitable remedies such as injunctions, as well as monetary damages, are available at Common Law.

Precedent is 'Judge-made Law'. The Judge's stated reasons for his decision (1) are binding on lower courts unless they can distinguish the case they are trying from the precedent in some way. However, observations which are part of the judgement but not central to it (2) are merely persuasive; lower courts do not have to follow such remarks if they are not found to be convincing.

Statute Law A Statute is an Act of Parliament. Some statutes contain **enabling** clauses which permit administrative orders to be made by the authority delegated by the statute to Ministers or Statutory Bodies (e.g. the Health and Safety Executive). Such Statutory Instruments are known as **Delegated** or **Subordinate** Legislation. They are of two types:-

(i) **Affirmative** S.I.'s which will not become law unless specifically approved by Parliament; and

(ii) **Negative** S.I.'s which automatically come into force after they have been laid before Parliament for forty days, unless a Member specifically asks for their annulment to be debated (3).

---

(1) His 'ratio decidendi'.
(2) 'Obiter dicta'.
(3) The Member is said to 'pray against' the Statutory Instrument.

Codification It sometimes happens that Statute Law and Common Law deal with the same subject; as a result, the legal situation can become somewhat tangled. Parliament may then pass legislation which not only collects the existing Common Law and Statute Law principles in a single statute, but may also add new legal principles to the law on a subject; for example the Sale of Goods Act 1893.

Consolidation Statutes and S.I.'s may amend what has gone before. Consolidation is merely a 'tidying up' operation; existing legal rules are gathered into a single statute, but no new legislation is passed. The Road Traffic Act 1972 consolidated earlier Road Traffic Acts, the Road Safety Act 1967, and certain Lighting Regulations.

11.1.2 The Courts

The Courts Act 1971 had the effect of considerably simplifying the rather complicated structure of the courts, making the first large-scale reorganisation of the judicial system since the Judicature Act 1873. Civil and criminal cases are dealt with separately.

A **Civil Action** lies between two parties, one of whom, the **plaintiff**, seeks a remedy (damages, injunction etc.) from the **defendant**. It is usual to refer to civil cases as Plaintiff (name) versus Defendant (name) followed by the date, e.g. Herrington v. British Railways Board 1971.

Criminal Actions are taken by the relevant prosecuting authority against a defendant who is being tried for a criminal offence, usually against statute law. Where the prosecutor is a representative of the State, e.g. the Police, the name is shown as Regina (The Queen) versus John Smith. Alternatively the prosecutor may be the Director of Public Prosecutions (DPP) or, less usually, the name of an officer, for example a local authority's Consumer Protection Officer who might be prosecuting a haulier whose vehicle is found to be overweight.

On **Appeal**, the defendant becomes the **respondent** and the plaintiff becomes the **appellant**, and the names are reported vice versa, e.g. Respondent v. Appellant 1977. Appeals may generally only be made on a point of law, not of fact (unless fresh evidence is available). The higher court accepts the facts determined by the lower court, and concerns itself only with the legal correctness of the lower court's decision.

The structure of the Criminal Courts is shown diagramatically below.

THE HOUSE OF LORDS

▲

(On Appeal)

▲

THE COURT OF APPEAL (CRIMINAL)

▲

(On Appeal)

▲

CROWN COURTS

▲ ▲

(On Appeal) (On Committal)

▲ ▲

MAGISTRATES' COURTS

The Magistrates' Court is described as a court of **summary jurisdiction** and may hear cases, typically traffic offences, and impose penalties (fines and imprisonment), as well as committing more serious cases to a Crown Court for trial by jury. Appeals from there are to the Court of Appeal, which may in difficult cases grant leave to appeal to the House of Lords.

Most **Civil Cases**, where damages for negligence claimed are less than £5,000, or breach of contract cases involving less than £5,000, are dealt with by County Courts. However, larger civil claims are dealt with by the Civil Division of the Crown Court (in which High Court judges may sit from time to time). Appellant procedure is similar to that described above. Whilst in theory a jury of 8 can be called, in practice the action is heard by a single judge assisted by registrars. Unless the parties object, a registrar may try actions where the claim is less than £500; or he may try any action with the parties' consent. Appeals from registrars' decisions can be made to a County Court judge.

Arbitration County Court Arbitration procedure introduced in 1973 ensures that small claims under £500 are automatically referred to a Registrar's arbitration (if either party requests he can refer the dispute to the judge or to an outside arbitrator). Like Tribunals, below, the advantages of informality, speed, subject expertise and cheapness are claimed for arbitration, at the expense of some small measure of 'rough justice' - strict rules of evidence, for example, do not apply, and the essential 'no cost' rule means that neither side can recover their legal costs. Arbitration is ideally suited to settling small business disputes of the type common in the transport industry. It has re-established County Courts as accessible 'small claims' courts.

Tribunals The Transport Manager, if he is involved in litigation, may well find his case dealt with by an administrative tribunal (4). In particular, the Traffic Commissioners are the Licensing Authority responsible for the grant, revocation etc. of an O-licence. Cases involving contracts of employment are heard by Industrial Tribunals, whose scope today is vastly increased since their inception in 1964 (5).

11.2 The Carrier and the Law

11.2.1 Common and Private Carriers

Anyone who carries for hire and reward incurs the obligations and liabilities of the carrier's profession. Unless the carrier stipulates the terms and conditions under which he carries, his customers are entitled to expect that he will take full responsibility for the goods accepted. In legal terms he
(a) does not limit his liability for negligence;
(b) is an 'insurer' of the goods;
(c) is a **Common Carrier**.

He may be a Common Carrier of certain goods or on certain routes if he agrees unconditionally to accept such traffic, and can only be sued for refusal to carry those goods of which he is a Common Carrier (6).

However, he may escape liability by relying on certain accepted exclusions:-
(i) Act of God;)
(ii) Act of the Queen's enemies;) These are both standard exclusion clauses.
(iii) Inherent vice of the consignment, i.e. some natural fault such as perishability or fragility;
(iv) Fault of the consignor, such as incorrect addressing or faulty packaging, provided that the carrier has done what he can to minimise the loss (7).

Private Carriers carry under individual contracts of carriage made with each customer. Carriers almost invariably draw to their customers' attention their

---

(4) Tribunals and Inquiry Act, 1958.
(5) Industrial Training Act, 1964.
(6) Johnson v. Midland Railway, 1849.
(7) Nugent v. Smith, 1872.

Conditions of Carriage, usually by inserting a statement on quotations, invoices and letterheads, specifically for the purpose of incorporating these into the contract, thereby making the carrier a Private Carrier. Thus Private Carriers are the rule and Common Carriers are the exception.

11.2.2 The Carriers' Act 1830

The Act was passed to limit the rather onerous Common Law liabilities (outlined above) of the Common Carrier, while at the same time prohibiting him from unilaterally imposing his conditions on all his customers by the simple device of a Public Notice (referred to in the Act as a 'Special Contract').

In essence the Act limited the carrier's liability for loss or damage to certain specified goods of high value but small bulk, to £10 per package unless:-
(i) the consignor declares the value of the goods and
(ii) agrees to pay any increased charge demanded by the carrier.

The carrier must post a notice in his receiving office detailing the increased charge, but he need not draw attention to the requirement to make a declaration of value. If no declaration is made, the carrier is exempt from any liability exceeding £10 per package.

11.2.3 The Contract of Carriage

A carrier, by virtue of the way in which he advertises himself as a carrier of goods, is in effect showing his willingness to enter into a contract with his customer. (In legal terms he is making 'an invitation to treat' (8)). However it is the consignor who makes the offer (that he has goods to consign) and the carrier, in accepting the offer for a consideration or payment (technically referred to as the 'freight') completes the contract.

In the event of a claim being made on the carrier by the owner of the goods (the plaintiff), the carrier (the defendant) has the task of proving that he has not been at fault, or negligent. The ownership of the goods passes from the consignor to the consignee at the moment when they are accepted by the carrier. From then on the consignor acts as the consignee's agent if he presses a claim for loss or damage.

11.2.4 Limitation of the Carrier's Liability

A Private Carrier by land is only liable for the negligent acts of himself or his servants. Even this liability may be too much for the carrier to shoulder, but if he attempts to exclude it through a clause in his conditions of carriage, he can only do so if he uses clearly unambiguous wording (9).

The Unfair Contracts Terms Act 1977 prevents the inclusion of terms excluding liability for negligence, unless these satisfy the requirements of reasonableness. Further, where such terms are incorporated in standard terms and conditions, the 'affected consumer' (defined as the party which, unlike the contractor making the standard terms, does not normally contract in the course of business) is protected from terms either
(a) restricting liability for breach of contract; or

---

(8) Wilkie v. London Passenger Transport Board, 1947. If a consignee overstamps the carrier's delivery note to the effect that the goods are received on the consignee's terms, this new condition becomes a prevailing counter-offer (Crutchley v. British Road Services, 1968; Denning J., Court of Appeal).
(9) Gillespie Bros. & Co Ltd. v. Roy Bowles Transport Ltd., 1973.

(b) offering an alternative entitlement to a substantially different or partial performance;

unless such terms also satisfy the requirements of reasonableness.

The onus of establishing the reasonableness of the term is on the party claiming this (10). Where (as is common in contracts for the carriage of goods) the term restricts liability to a specified sum of money and the question of the reasonableness of such a term arises, courts will have regard to the resources that the party might be expected to have ready to meet such liabilities, and to the extent to which it was open to him to insure his liability.

Delay The carrier is held liable if loss or damage occurs to his customer through his unreasonable delay, and if such loss was reasonably foreseeable as a direct consequence of the delay (11).

Delivery The carrier's liability in any case ceases when transit ends, i.e. when delivery is effected. If goods are **wrongly delivered** through the **wilful misconduct** of the carrier, then even conditions of carriage excluding liability for negligence will clearly not protect him (12). However, such conditions will protect him if he or his servant merely **negligently misdelivers** the goods (13).

Bailment A warehouseman, who accepts goods for storage in exchange for payment, is a 'bailee for reward', i.e. he is a custodian of the goods of his customers and has a strict duty to take care towards them (14). However, a carrier who is unable to make a delivery because of some fault of the consignor or consignee is effectively a 'gratuitous bailee', i.e. he will not be paid for storing the goods. In such circumstances his only duty is to avoid negligence (15). Note that if under his conditions he is able to charge for **demurrage**, this is no more than an attempt to recover from his customer some compensation for the loss of use of his vehicle. However, he may detain the goods on his vehicle or at his premises in order to recover his freight charges.

Lien The legal right of **lien** enables a tradesman, such as a carrier, whose customer's goods are lodged with him e.g. for transport, storage or processing, to retain these goods until the customer pays his charges. All carriers have in law a **particular lien** enabling them to retain any particular consignment until it is paid for. They can extend this in their conditions of carriage to a **general lien**, which will enable them to detain any goods consigned by a customer who owes money in respect of any other consignment. The carrier cannot charge for warehousing the goods in order to enforce a lien (16), and the right of lien does not confer the right to sell the goods. Any unauthorised sale will result in loss of lien following loss of possession; the carrier may also be liable to the customer for the civil wrong or **tort** of conversion.

Dangerous goods There can be no such thing as a Common Carrier of dangerous goods, since every carrier is always free to refuse to accept them for carriage (17). If dangerous goods are sent without the knowledge of the carrier, the

---

(10) Contracts may also be referred to the Office of Fair Trading. In November 1979 the Office of Fair Trading asked the Road Haulage Association to reconsider some time clauses in their Conditions of Carriage.

(11) Hadley v. Baxendale, 1854.

(12) Hoare v. Great Western Railway, 1877, and a related case concerning bailment, Alexandre v. Railway Executive, 1951.

(13) Stephens v. Great Western Railway, 1885.

(14) Coggs v. Bernhard, 1704.

(15) Stephenson v. Hart, 1828.

(16) Somes v. British & Empire Shipping Co., 1860.

(17) Barnfield v. Goole & Sheffield Transport, 1910.

consignor is liable; otherwise the carrier will determine the conditions under which he will accept them and the precautions which he needs to take.

Stoppage 'in transitu' If the vendor of goods (by definition the consignor) learns of the inability of the buyer to pay for his consignment due to the latter's insolvency (not necessarily bankruptcy), he may instruct the carrier to stop the goods in transit and return them to him, naturally at the consignor's expense (18). This right arises in these circumstances irrespective of the fact that ownership of the goods at the time of stoppage will have passed to the consignee (the buyer).

11.2.5 Negligence

Negligence amounts to a failure to exercise the ordinary duty of care which the carrier owes to his customer, or which for that matter any person owes to any other person who, he could reasonably have foreseen, might suffer injury or damage by his actions (19). Suppose for example an accident occurred as a result of the carrier's breach of his statutory duty of care in not maintaining his vehicle; clearly the carrier is negligent (20). He must also exercise reasonable care for the safety of all visitors to the premises which he occupies; this includes persons either invited or licensed to call there (21).

Trespass, the unlawful interference with a person's possession of land, is actionable even if the plaintiff suffers no damage as a result. Whilst an occupier does not owe the same duty of care to a trespasser as to a visitor, liability can still arise where there is such probability that the trespasser will be exposed to danger that by standards of common sense and common humanity the occupier could be said to be to blame for not taking reasonable care.

A distinction must be made between premises where active operations are carried on, and those where this is not so. In the former case a higher duty of care rests upon the carrier, in particular where the presence of trespassers might reasonably be foreseen. Thus failure to repair a fence around a depot situated next to an area where young people play, would amount to negligence (22).

Contributory negligence The law distinguishes a special case where the defendant in a negligence case, by virtue of his own acts or omissions, has contributed to the loss which he has sustained (23). In such cases damages are apportioned between the parties in proportion to their respective degrees of negligence (24).

Burden of proof of negligence The burden of proof in negligence cases is usually on the plaintiff who makes the claim. However, in some cases this rule might be unfair to him, for example where it seems obvious on the face of it that the defendant has been negligent. In such cases the burden of proof can be reversed by making the assumption that the defendant's negligence must have been the cause of the damage sustained, unless the defendant can show otherwise (25).

Employer's liability Not only will an employee be responsible in law for his own actions; in some cases the employer may also be liable for the employee's

---

(18) Sale of Goods Act, 1893, section 44.
(19) Donoghue v. Stephenson, 1932.
(20) Barkway v. South Wales Transport, 1950.
(21) Occupier's Liability Act, 1957, and Health and Safety at Work etc. Act, 1974; see also Chapter 12, section 12.5.
(22) Herrington v. British Railways Board, 1971.
(23) Harvey v. Road Haulage Executive, 1945.
(24) Law Reform (Contributory Negligence) Act, 1945.
(25) 'Res ipsa loquitur' (The thing speaks for itself).

actions, in particular where there is a contract of employment and the employee is carrying out those duties for which he was employed (26). 'Agency' drivers pose a special problem, as they are the servants of their agency and are not employed by the carrier who hires them; nevertheless their activities are under the control of the hirer. The carrier's responsibility for them would certainly extend to ensuring that they comply with relevant statute law (27). Situations sometimes arise where drivers give lifts or engage in some other 'frolic of their own', and where the employer, knowing this, attempts to restrict his liability by, for example, displaying a 'Cab Notice' disclaiming responsibility for unauthorised passengers. However, this in itself is insufficient; it must be shown that the passenger was made aware of the restriction or could reasonably have been expected to have known of it, and nevertheless still consented to accept the risks involved (28). Much the same principle is involved in deciding whether any person in contract (as opposed to tort, above) can be considered as having consented to another person limiting his liability to him. A whole series of 'ticket cases' have been developed to clarify this point. In general, only if the defendant can show that the plaintiff was positively made aware of the limitations which the defendant intended to introduce, can a defence based upon a notice or ticket be of any use (29), and then only if the limitation satisfies the test of reasonableness (30).

Nuisance involves inconvenience to the public or to an individual or both. Employers can be vicariously liable for their employees' actions.

Public nuisance is an act which increasingly interferes with the public at large. Whilst it is a **crime**, an individual can still bring a civil action if he can show damage over and above that suffered by the public at large. For example, a vehicle inconsiderately parked may be a public nuisance (regardless of any statutory restrictions), but if it blocks access to an individual's premises, that individual may be able to demonstrate actual financial loss. **Private nuisance**, on the other hand, gives rise only to civil liability. It is defined as unlawful interference with a person's use and enjoyment of his land. Remedies provided by the courts include restraining **injunctions** or the award of actual **damages**.

11.3 Company Law

Many haulage undertakings are registered as Limited Companies (31). However, haulage is a very fragmented industry with a high proportion of owner-drivers (who as O-licence holders will require to be professionally competent and of good repute) and small firms. In addition a large sector of the industry is controlled through Public Limited Companies belonging to the National Freight Corporation or the Transport Development Group. Thus there is found a full spectrum of types of ownership in the industry.

11.3.1 The Sole Proprietor

The sole proprietor, trading on his own account, financing his activities from his own funds, bank loans, hire-purchase etc., and accepting absolute personal liability for the financial affairs of his business, is often an owner-driver or small employer.

---

(26) Limpus v. London General Omnibus Company, 1862.
(27) Ready Mixed Concrete (East Midlands) v. Yorkshire Area Traffic Commissioners, Transport Tribunal 1970.
(28) Bennett v. Tugwell, 1971.
(29) Thornton v. Shoe Lane Parking, 1971.
(30) Unfair Contracts Terms Act, 1977.
(31) Companies Acts, 1947, 1968 and 1980.

11.3.2 Partnership

The sole proprietor may expand his business and its financial base by going into partnership (32) with up to six other partners (totalling seven), who will invest their capital and expertise. However, the drawback of unlimited personal liability, although it is a shared burden, still persists. Any of the partners individually, or all of them jointly, can be made liable for the whole debts of the business. Alternatively, the partnership can be a limited partnership (33) of up to 20 persons. In such a firm, the limited partners simply invest their capital but do not control the business (34), and are liable only to the amount invested; but at least one general partner must still assume unlimited liability for the affairs of the business.

11.3.3 Limited Companies

These are companies in which the liability of the members to meet the company's debts on insolvency is limited - either by share or by guarantee. They are corporate bodies registered by the Registrar of Companies (31), after a process known as **incorporation** which involves the submission of a **memorandum of association** and **articles of agreement**.

A Private Company may commence business as soon as its Certificate of Incorporation is granted, but a Public Limited Company (PLC) may not - in particular, it may not exercise any borrowing powers until the Registrar of Companies certifies that it has complied with the requirements as to share capital.

The Articles of Association regulate the internal administration of the Company. A private company must have a Company Secretary and at least one Director; a PLC must have at least two Directors (one of whom may be the Company Secretary).

Requirements for registration of PLCs are stringent; there are restrictions on dealings between Companies and their Directors, in particular property transactions and employment and service contracts. Insider dealing (i.e. the improper use of boardroom information for personal gain) is a criminal offence.

The Memorandum of Association shows:-
(i) The name of the Company;
(ii) The address of the registered office;
(iii) The aims and objectives of the Company. This is important since the question of the Company's activities may subsequently arise. These activities must in all cases be within the provisions of the Memorandum (31)(36). The memorandum should state if the company is to be registered as a PLC (37);
(iv) The authorised share capital, i.e. the **maximum** amount which the Company may raise by share subscription, as opposed to the amount of capital currently called up by the Directors (38);
(v) A statement that the liability of the members is limited;

---

(31) Companies Acts, 1947, 1968 and 1980.
(32) Partnership Act, 1890.
(33) Limited Partnership Act, 1907.
(34) They are said to be 'sleeping partners'.
(35) Companies Act, 1980.
(36) The 'ultra vires' (beyond the powers) rule.
(37) Private Limited Companies could re-register as PLCs (if they met the criteria) between 22nd December 1980 and 22nd March 1982. Any new company limited by guarantee (i.e. having no shares) must by definition be a Private Limited Company.
(38) i.e. the 'subscribed capital'.

(vi) A signed undertaking by the members to purchase shares against their names.

Control of the Company is vested in a Board of Directors elected by shareholders, and the Company's internal affairs are regulated by its Articles of Agreement.

Private Limited Companies may have two or more members (there is no upper limit), all of whom must hold shares. These shares cannot be bought or sold on the open market; indeed, it is a criminal offence to do so. Any body (other than a PLC or Private Limited Company) trading as a registered company will have to have a name ending in 'unlimited'. However, some Charities and Trade Associations limited by guarantee are excluded (35).

Public Limited Companies In the main, these are those Public Companies quoted on the Stock Exchange, where their shares can be bought and sold. This permits such Companies to raise large sums of capital. This is the preferred form of organisation of large haulage companies, for example those of the Transport Development Group. The registered names of public companies must now end in the words 'Public Limited Company' (PLC) (35). A minimum of two persons may now form a PLC, but there is a new requirement to include an issued share capital of at least the authorised minimum (currently £50,000) (35).

The Registrar of Companies keeps an index of names of registered firms, including those previously registered with the Registrar of Business Names.

The Companies Act 1981 contains provisions relating to the audited accounts of Companies (39). Small Companies (40) need cnly file an abbreviated balance sheet with the Registrar of Companies. Medium-sized Companies (41) must file a full balance sheet and Directors' Report, but need not draw up their Profit and Loss Account so as to disclose turnover. All other Companies, all PLCs and all Insurance Companies must file full accounts. Company accounts can be inspected by the public on payment of a small fee.

Capital of a Company is derived from the issue of shares. **Ordinary shares** carry voting rights, and attract a dividend where this is declared by the Directors. **Preference shares** entitle the holder to a fixed percentage return before all other claims on the profits. They often carry no voting rights. **Debentures** also are non-voting and entitle their holders to a fixed return, but are not part of the Company's share structure (i.e. not part of the authorised or subscribed capital). They are in fact loans made by investors against the security of the assets (or of a particular asset, typically a vehicle), and debenture holders in the event of the liquidation of a Company have first call on the assets.

Liquidation is the legal process by which a Company is closed down ('wound up'), its assets realised and its creditors paid off. A minority of shareholders may petition the courts for a winding-up order (35). Liquidation is a more drastic procedure than the appointment of a Receiver, who runs the business for the Directors, to attempt to repay the preferential creditors.

---

(35) Companies Act, 1980.
(39) The Act applies the EEC's Fourth Directive on Company Law.
(40) Those with turnover less than £1.4 million and balance sheet total less than £0.7 million.
(41) Those with turnover less than £5.75 million and balance sheet total less than £2.8 million.

12 Employment Law

12.1 Industrial Relations

The Transport Manager will necessarily be continually concerned with the creation of good industrial relations at the workplace. Relations between employer and employee may be conducted both at the individual level, and also by collective bargaining. Modern employment law is concerned not only with creating a 'floor of rights' (1) for the individual employee, but also with the establishment of machinery for promoting improvements in industrial relations (2). The next few sections which deal with these points are of especial relevance to all employers and to Transport Managers in particular.

12.1.1 The Trade Union and Labour Relations Acts 1974/1976

(As amended by the Employment Act 1980)

These Acts repealed the Industrial Relations Act 1971, but it should not be thought that none of the 1971 Act's provisions survived in any way. On the contrary, the 1974 Act strengthened the law relating to unfair dismissal (see below), and restored to all 'independent trade unions' (simply defined in the Act as 'organisations of workers') the immunities which all of them (3) had enjoyed from civil (4) and criminal (5) proceedings before 1971. The right to belong (6) to a trade union is central to the Act, which provides that where there exists a Union Membership Agreement, it will not normally be unfair to dismiss an employee who refuses to belong to an independent trade union specified in an 'approved' agreement, unless such a refusal is based on grounds of conscience or other deeply held personal convictions. Additional exemptions from 'closed shop' membership are added by the Employment Acts 1980 (7) and 1982 (8).

---

(1) Wedderburn, 'British Journal of Industrial Relations', vol. X no. 2, 1972.
(2) Employment Protection Act 1975.
(3) Whether or not they registered under the 1971 Industrial Relations Act.
(4) Trade Disputes Acts 1906 and 1965. Now Trade Union and Labour Relations Acts 1976, section 13.
(5) Conspiracy and Protection of Property Act 1875, and Criminal Law Act 1977.
(6) Originally Industrial Relations Act 1971, section 5. However, there is now no automatic right not to belong to a Trade Union, although Union Membership Agreements cannot be retrospective or bind dissenters who have never belonged (Employment Act 1980).
(7) Employment Act 1980 adds new sections 3a to 3e to Employment Protection (Consolidation) Act 1978, widening the exemptions from Union Membership Agreements (see figure 25).
(8) Provided for agreements to be periodically reviewed and confirmed by ballots; increased compensation, including retrospective compensation, for unlawful dismissal after a Union Membership Agreement; and allowed Trade Unions to be 'joined' in actions. See 12.2.4 for a full treatment of these points.

The Trade Union and Labour Relations Acts also

(a) Provided that collective agreements should not normally be legally enforceable unless they contain a specific written clause to that effect;
(b) Abolished the National Industrial Relations Court, the Commissional on Industrial Relations and the Registrar of Trade Unions and Employers' Associations;
(c) Restored the pre-1971 rights for all (3) 'independent trade unions' to take such sanctions as strikes and peaceful picketing, and prevented employers from obtaining 'ex parte' (9) injunctions in a trade dispute.

These rights are now subject to the following limitations:-

Sanctions taken 'in furtherance of a trade dispute' cannot give rise to criminal or civil action provided that the sanction itself (if taken by an individual) is not unlawful. **Pickets** are only lawful in these circumstances (10) if they attend peacefully at or near their own place of work (11). Immunity from civil actions for breaches of a contract continue in force (4) where **primary and direct secondary action** (12) is taken, but employees involved in remote or indirect action lose this immunity when they interfere with commercial contracts. Secondary action which goes beyond the following limits is deemed to be too **remote** to create any civil immunities. The tests of what constitutes direct secondary action are:-

(a) Motive. Its principal purpose must be to prevent or disrupt supplies or goods which might substitute for those under a contract between an employer (or an associated employer who is a party to the dispute) and his first supplier or first customer.

(b) Capability. The action must be likely to achieve this purpose (13), and, in addition, where picketing takes place this must be lawful (as defined above) and carried out by employees, or their full-time union officials, who are a party to the dispute. (See Figure 22).

12.1.2 The Employment Protection Act 1975 and the Employment Protection (Consolidation) Act 1978 (As amended by the Employment Act 1980)

These are wide-ranging Acts creating not only considerable individual rights, and - as their titles imply - improving employees' security of employment, but also creating a whole new legal infrastructure for promoting improved industrial relations. This latter provision is all that remains in the original 1975 Act.

---

(3) Whether or not they registered under the 1971 Industrial Relations Act.
(4) Trade Disputes Acts 1906 and 1965. Now Trade Union and Labour Relations Acts 1976, section 13.
(9) e.g. an injunction obtained by an employer restraining his employees' Union from industrial action (such as 'blacking') without the presence in court of the defendant or his representative.
(10) The Employment Act 1980, section 16, inserts a new section 15 into the Trade Union and Labour Relations Act for this purpose.
(11) Unemployed workers who have lost their jobs in the trade dispute may picket their former place of work, and workers with more than one workplace may picket the premises of their employer from which they work or from which their work is administered. Code of Practice on Picketing, section 13; see also footnote 54.
(12) Secondary action is action taken against an employer not a party to the trade dispute. Employment Act 1980, section 17.
(13) Thus reversing the House of Lords decision in McShane v. Express Newspapers, 1980.

Fig. 22 TRADE UNION IMMUNITIES

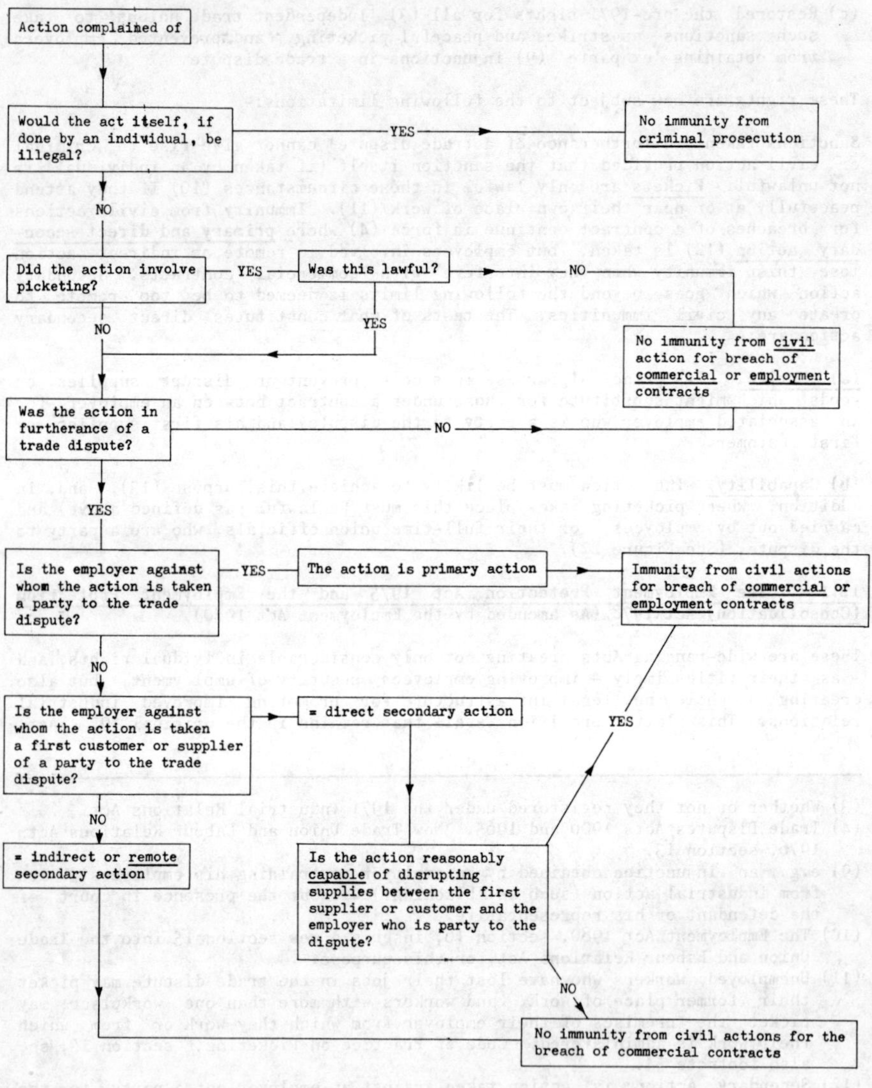

Fig. 23 DISCLOSURE OF INFORMATION

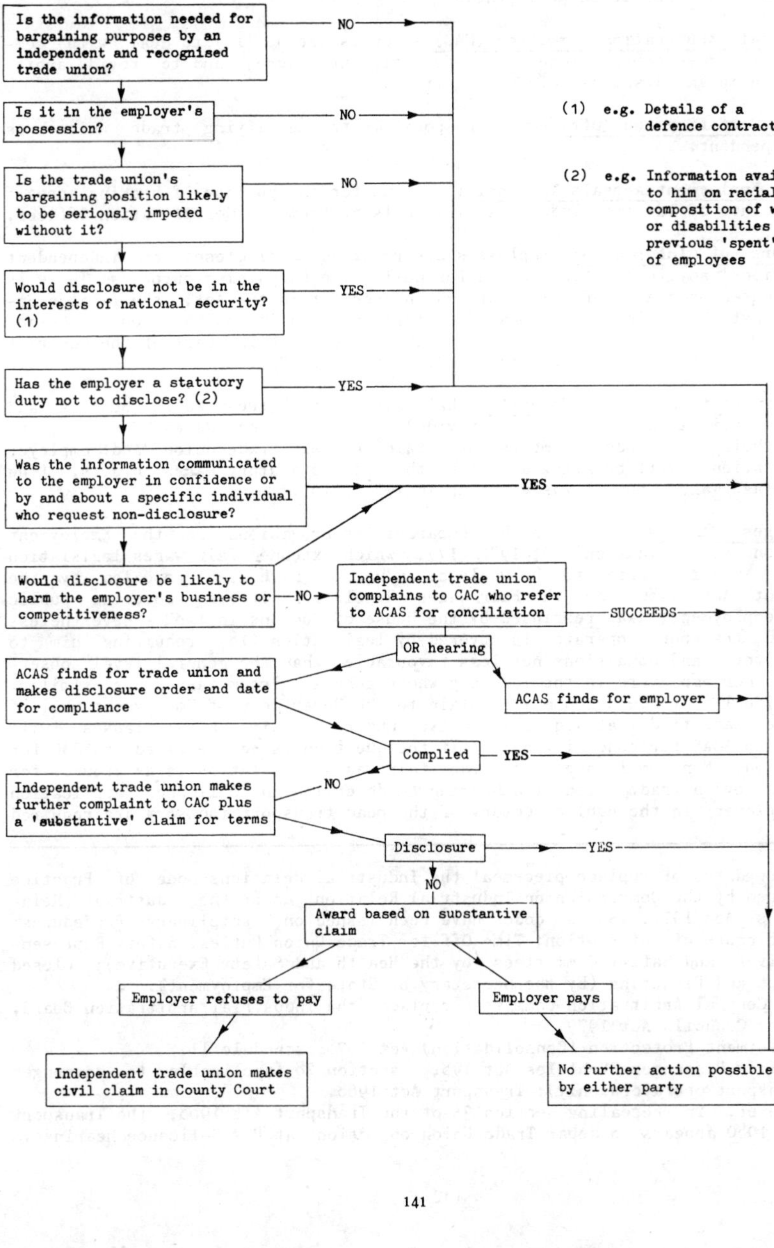

(1) e.g. Details of a defence contract

(2) e.g. Information available to him on racial composition of workforce or disabilities or previous 'spent' convictions of employees

Institutions The following new institutions were set up under the Act:-

(a) **Advisory, Conciliation and Arbitration Service (ACAS)** is a statutory body charged with the responsibility of offering advice, assistance and conciliation to employers and employees alike; inquiring into specific industrial relations questions; issuing Codes of Practice (14); and referring matters in dispute to Arbitration or to the:-

(b) **Central Arbitration Committee (CAC)** which is set up (15) to deal with complaints under the disclosure of information procedure, and to provide arbitration in any disputes referred to it.

(c) **The Certification Officer** is responsible for certifying trade unions as independent.

(d) **The Employment Appeals Tribunal** is set up for the purpose of hearing appeals on points of law from Industrial Tribunals or from the Certification Officer.

Disclosure of Information Employers are required to disclose to 'independent trade unions' any information needed for collective bargaining purposes. There is a complicated procedure of complaint and enforcement in the Act; the final sanction against the employer is an award of terms and conditions based on a substantive claim by the union, which then has the effect of being part of the Contract of Employment of the employees concerned. (See Figure 23).

Statutory Joint Industrial Councils, half-way houses between Wages Councils and free collective bargaining, are provided for. In essence Wages Councils, by losing their independent members but retaining their trade union and employer representation, will be allowed to make their own statutory wages orders. (The Road Haulage Wages Council was wound up in 1978) (16).

Fair Wages Clauses Although the comparability provisions in the Employment Protection (Consolidation) Act 1978 (17), which extended fair wages legislation from the public sector to the whole of industry, have been repealed by the Employment Act 1980 - and the Fair Wages Resolution itself, covering public sector employment, was rescinded by the House of Commons in 1982 - nevertheless the road transport operator is covered by legislation (18) requiring him to observe 'terms and conditions not less favourable' than the general level observed by other employers in the industry whose general circumstances are similar. An individual or trade union may complain to the Department of Employment who, if they are satisfied that a question arises under the fair wages clauses, will refer it to ACAS for conciliation or, if the question is not resolved, to CAC for arbitration. Non-compliance with the fair wages legislation is a ground for objection by a trade union at a hearing to determine an O-licence application (19). Employers in the public sectors of the road transport industry are required

---

(14) To update or replace piecemeal the Industrial Relations Code of Practice issued by the Commission on Industrial Relations under the Industrial Relations Act 1971. So far codes have been issued on Disciplinary Procedures; Disclosure of Information; Time Off for Trade Union Duties; Safety Representatives and Safety Committees (by the Health and Safety Executive); Closed Shops and Picketing (by the Secretary of State for Employment).
(15) The Central Arbitration Committee replaces the Industrial Arbitration Board.
(16) Wages Councils Act 1979.
(17) Employment Protection (Consolidation) Act 1978, schedule 11.
(18) Public Passenger Vehicles Act 1981, section 28 (applicable to passenger transport operators only); Transport Act 1968.
(19) However, in repealing section 35 of the Transport Act 1968, the Transport Act 1980 appears to debar Trade Union objections at PSV O-licence hearings.

(20) to establish machinery for collective bargaining with their employees' representatives.

Rights of Employees under the Act The following are basic rights of all employees. They are now all consolidated into the 1978 Act.

Guaranteed payments for a limited period for employees who would otherwise lose pay due to short-time working or layoffs (21);

Payment of normal wages during a period of **medical suspension** (22);

Maternity pay and rights It is unfair to dismiss a woman employee because she is pregnant. In addition she is entitled to be paid for the first six weeks of her absence, and may return to work at any time up to 29 weeks after her baby is born, subject to certain conditions qualified in most cases by what is reasonably practical (See Figure 24);

Trade Union membership and activities The Act gives every employee immunity from disciplinary action in respect of his activities in connection with a trade union (23). An individual has the right not to be **dismissed**, or be subject to **action short of dismissal** (24), because of trade union membership or activities, whether or not the union is independent. Equally, such action cannot be taken against an employee for the purpose of compelling trade union membership.

In the latter case, where an employer faces a complaint to an Industrial Tribunal of 'unfair dismissal' or 'action short of dismissal', he may, before the hearing, **'join'** (25) in the proceedings a person or trade union who he claims induced him, by actual or threatened industrial action, to act against the employee in the manner complained of for not being a member of a trade union. If the employee's complaint is upheld, the Tribunal may require the person or trade union so joined to **contribute** all or part of the compensation awarded (26). In the same way the employer may join another employer in the proceedings if these arose from a 'union labour only' term in a contract, and the second employer may in turn join another person or trade union as above.

Any industrial action taken to compel Trade Union membership (coercive

---

(20) Transport Act 1968, relating to the National Bus Company, Scottish Transport Group and Passenger Transport Executives.
(21) £9.75 per day maximum, for not more than 5 days in any 12-week period (Employment Act 1980, section 14(1)).
(22) Suspension from work under statutory regulations (e.g. The Radioactive Substances (Road Transport Workers) (GB) (Amendment) Regulations 1975) after medical examination.
(23) Employment Protection (Consolidation) Act 1978. He may, if he is dismissed wholly or partly for this reason, seek **interim relief** from an Industrial Tribunal pending determination of his complaint. Employment Act 1980, section 7, qualifies these rights where a Union Membership Agreement exists.
(24) 'Action short of dismissal' is not defined in the Employment Act 1980 section 15 (which amends sections 23-26 of the Employment Protection (Consolidation) Act 1978), and it is for the Tribunal to decide if it has occurred. It might cover discrimination in promotion, transfers or opportunities for training, or threats of dismissal or redundancy.
(25) Employment Act 1980, section 10, inserts new sections 76a/76b into the Employment Protection (Consolidation) Act 1978 for this purpose.
(26) Compensation for 'action short of dismissal' is similar to that for 'unfair dismissal' and may include e.g. a recommendation for re-instatement or re-engagement, but in addition the Tribunal will make a declaration to the effect that the employee's rights have been infringed.

Fig. 24 MATERNITY RIGHTS

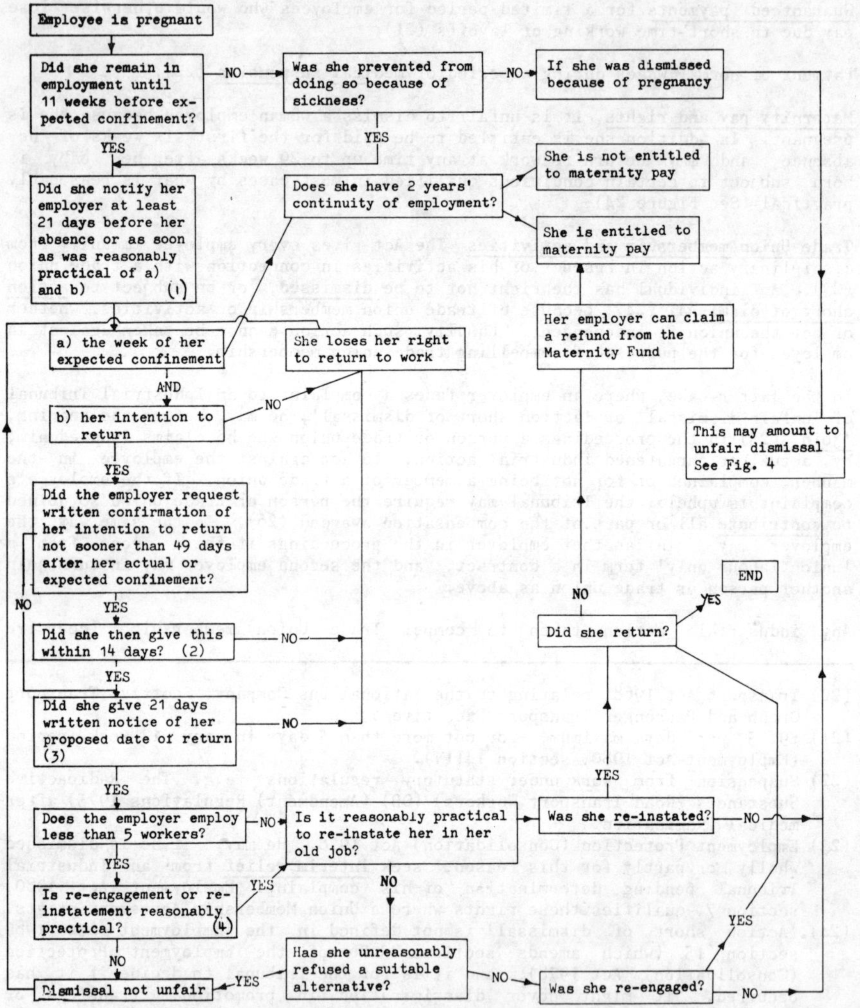

NOTES

(1) The notification must be in writing if the employer requests this.
(2) Extended to 28 days if she is prevented from doing so by her illness or by an interruption to her employer's business.
(3) She is only entitled to 29 weeks' maternity leave after the confinement, but she, or the employer, both have the right to postpone her return by not more than 4 weeks providing, respectively, either that a doctor certifies her unfit to return to work, or the employer notifies her of his reasons.
(4) The reason must be some reason other than redundancy, e.g. re-organisation.

recruitment), where the worker being forced to join the union works for a different employer than the person taking the action, and whereby a breach of a Contract of Employment or a commercial contract is caused, is no longer immune from civil action (27).

Protection against unreasonable expulsion or exclusion from a trade union Individuals who work, or seek employment, where a closed shop operates have the right not to be unreasonably (28) excluded from or expelled by the union. They may, within six months of any such action, complain to an Industrial Tribunal, which may make a declaration if it finds for the individual that he or she should be admitted. Where such a declaration has been made, the complainant may, between 1 and 6 months afterwards, apply for compensation from the tribunal if re-admission has taken place, or from the Employment Appeals Tribunal if not re-admitted.

An itemised Pay Statement;

A statutory period of notice of dismissal from the employer (see 'Contracts of Employment' below);

A written statement of the reason for dismissal;

Time Off. Officials of a recognised independent trade union are entitled to reasonable time off with pay during working hours to carry out their duties and to train. Members are entitled to similar time off, but not with pay (29).

Employees are also entitled to reasonable time off to carry out certain public duties.

An employee with at least two years' continuity of employment who is being made redundant, is entitled to reasonable time off with pay to look for work or to arrange training (30).

A woman who is pregnant is entitled not to be refused reasonable time off work with pay to receive ante-natal care.

Safety representatives are entitled to time off with pay to carry out their functions and attend training (31).

Protection from Unfair Dismissal The original provisions relating to Unfair Dismissal in the Industrial Relations Act 1971 were progressively strengthened by the Trade Union and Labour Relations Act 1974 and by the Employment Protection (Consolidation) Act 1978.

Claims by an employee to have been unfairly dismissed (as well as claims relating to the other employees' rights listed above) are disposed of by an Industrial Tribunal, which must consider the question having regard to the substantial merits of the case and the size and administrative resources of the employer

---

(27) Employment Act 1980, section 18, excludes coercive recruitment from the immunity under the old Trade Union and Labour Relations Act 1976 section 13.
(28) Employment Act 1980, section 4. The fact that a union's actions conform to its rules is not to be treated as conclusive evidence of their reasonableness.
(29) Code of Practice on 'Time Off for Trade Union Membership and Activities'.
(30) Compensation of up to two-fifths of a week's pay can be awarded by an Industrial Tribunal to an employee who is not allowed this time off.
(31) Safety Representatives and Safety Committees Regulations 1976, section 4(2) and Schedule.

(32). It must be satisfied that the reason for dismissal was fair, i.e. that it was because of the applicant's
(a) Redundancy, and that the method of selection for redundancy was not unfair;
(b) Incapability;
(c) Misconduct;
(d) Unlawfulness (for example, a driver may have lost his driving licence for an offence, and can no longer be employed as a driver);
(e) Refusal (on other than conscientious grounds) to belong to a union with which the employer has a union membership agreement (but see 12.1.1, also Fig.25).

The employee must actually have been dismissed, with or without notice. However, if the employer so contrives matters that the employee had no option but to resign, the Tribunal will construe this as constructive dismissal. Obviously, for there to have been a dismissal in the first place, the employee must not be in a category of employment which excludes the requirement for a Contract of Employment (see below). He must also, at the time of making the claim, have at least 52 weeks' continuity of employment (33), and have made the claim within 3 months of his dismissal, unless the Tribunal can be shown good reason for delay in claiming. If he is employed under a fixed-term contract for one year or more, his statutory dismissal rights may be waived if he has agreed in writing to exclude any claim for Unfair Dismissal, where the dismissal consists only of the expiry without renewal of the fixed-term contract.

If the tribunal finds that the applicant was unfairly dismissed, and the applicant wishes to be re-employed by the respondent, the Tribunal may make an order for reinstatement or re-engagement on not less favourable terms. Otherwise the applicant will be given an award of compensation (34).

<u>Job Security</u> The job security provisions in the Act mainly concern redundancy, and are dealt with below.

In addition there are a number of miscellaneous provisions in the Act. The Secretary of State is empowered to license private employment agencies (of the kind which provide agency drivers). The appointment of statutory safety officers (35) is now vested in recognised trade unions. Employees who work for employers who have less than four employees are no longer debarred from claiming unfair dismissal (33) (36).

(See Figures 25, 26, 27 and 28).

---

(32) The Tribunal may on application from one party, or of its own volition, hold a pre-hearing assessment of a claim and, if it considers the application unlikely to succeed, warn either party that, in such an event, if the claim is persisted with, an order for costs may be made against the party. Tribunal members who sit for such hearings are debarred from sitting at the full hearing. Tribunals (Rules of Procedure) Regulations 1980, section 61.
(33) If the employer has less than 20 employees, the qualifying period before claiming unfair dismissal is 104 weeks (Employment Act 1980).
(34) Composed of a basic award (based on the Redundancy Payments scale maximum 30 weeks at £140 per week (January 1983) = £4200), plus a compensatory award (maximum £7280 for 52 weeks), and up to the same amount again if the employer fails to comply with a reinstatement or re-engagement order. Greatly increased compensation is available to those unfairly dismissed as a result of a Union Membership Agreement.
(35) Health and Safety at Work etc. Act 1974.
(36) Employment Protection (Consolidation) Act 1978, schedule 16, paragraph 4.

FIG. 25 DISMISSALS AS A RESULT OF UNION MEMBERSHIP AGREEMENTS

Employment Protection (Consolidation) Act 1978, section 58
(As amended by the Employment Act 1982, sections 3 and 10)

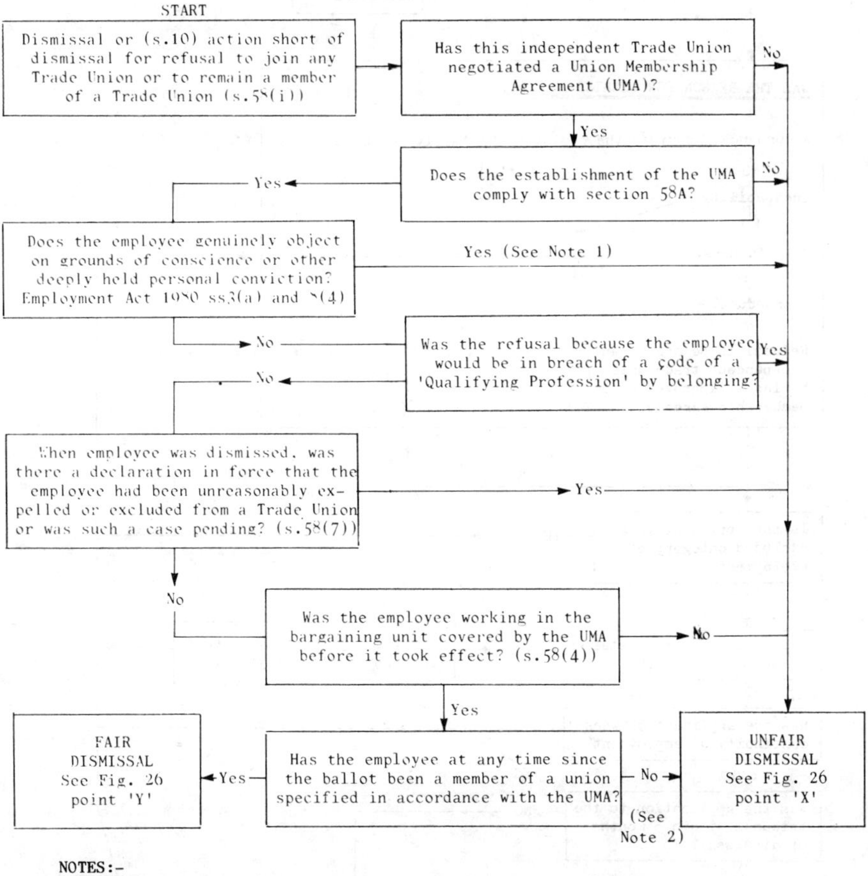

NOTES:-

(1) Thus re-establishing the right not to belong to a Trade Union.

(2) This bans retrospective UMAs, and allows existing members of non-signatory unions to retain such membership, and members of the signatory union to resign before the UMA becomes effective. It thus provides an opportunity for a dissenting minority of existing employees not to have to abide by the majority decision - whether the dissenters are non-union members, or members of a non-signatory union.

Fig. 26 UNFAIR DISMISSAL

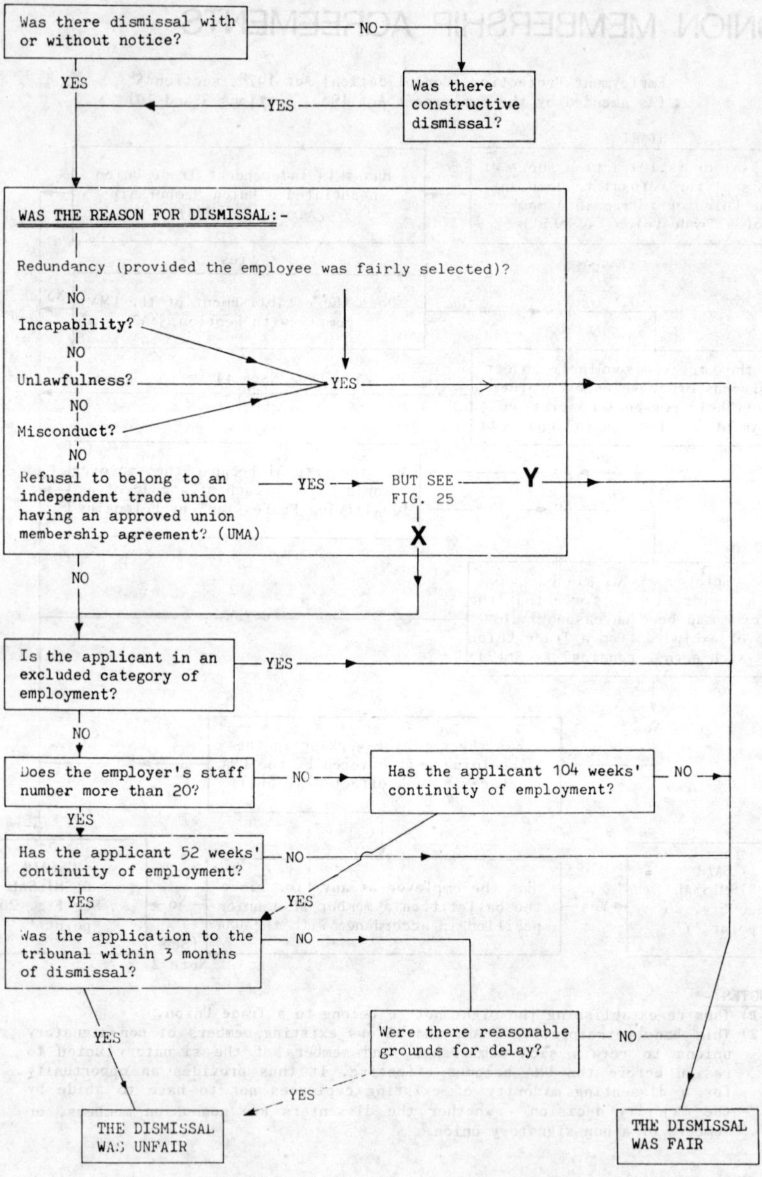

FIG.27 REMEDIES FOR UNFAIR DISMISSAL

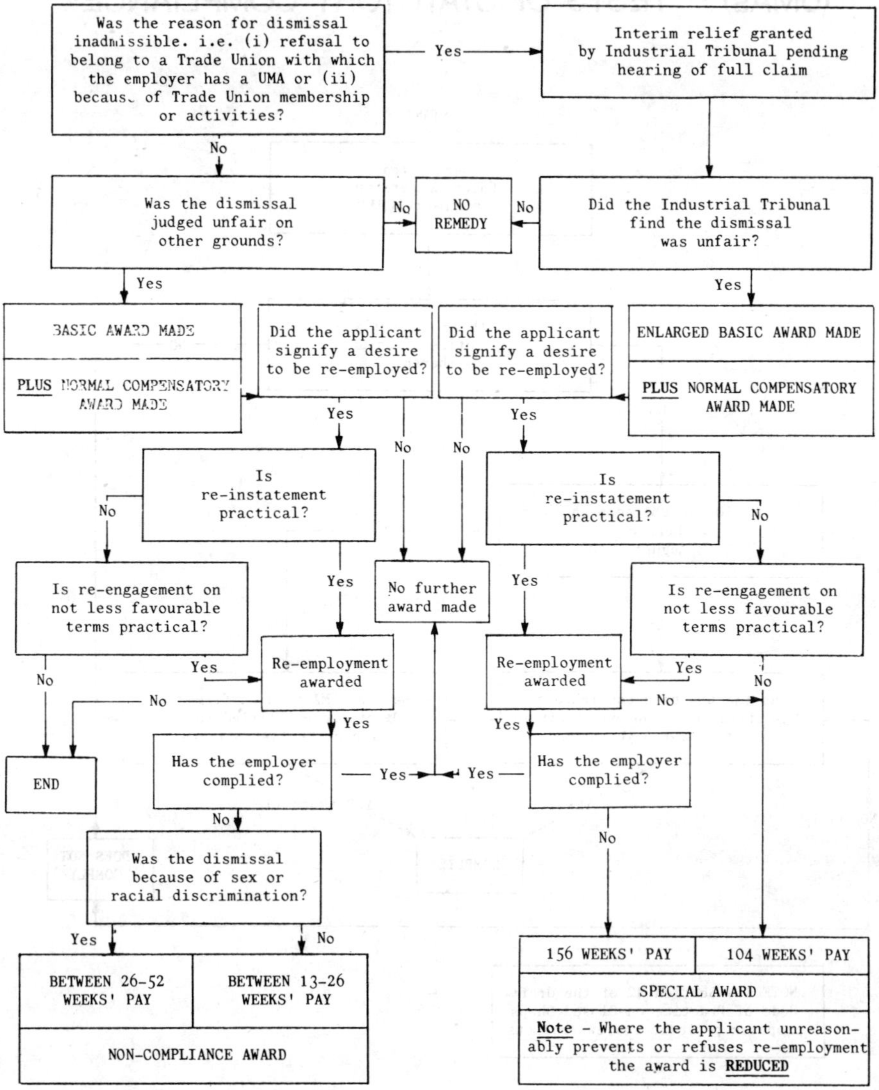

FIG. 28 UNION MEMBERSHIP AGREEMENTS (UMAs) — TESTS OF STATUTORY COMPLIANCE

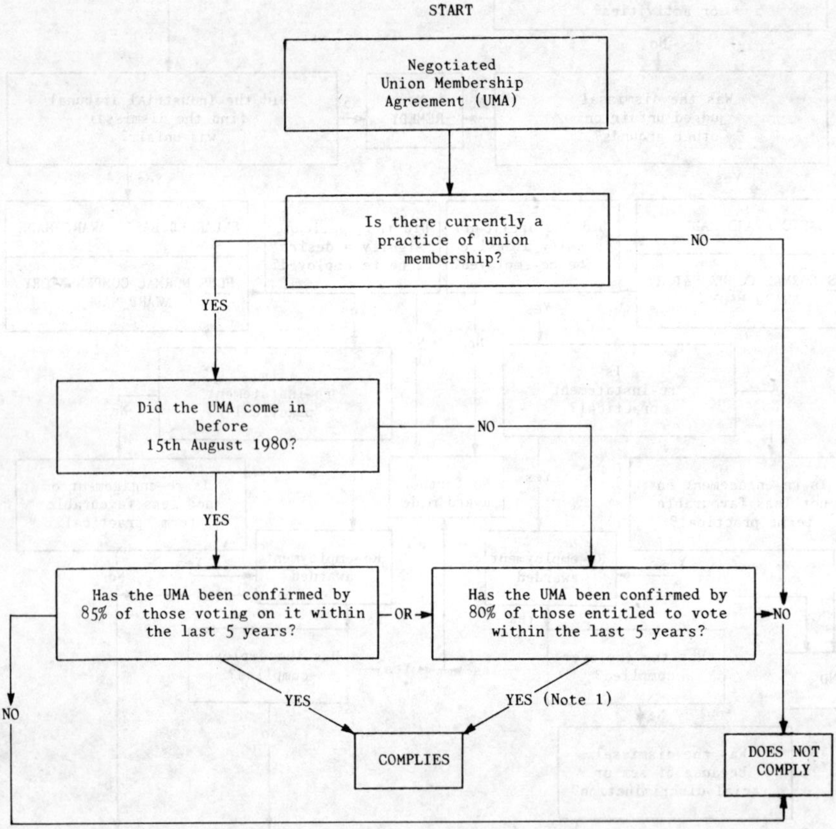

NOTE 1 - Section 42 of the draft Code of Practice on Closed Shops proposes periodic reviews/ballots

Contracts of Employment

Every full-time employee, after 13 weeks' employment, is entitled to a written statement of his terms and conditions of employment. The statement can be comprehensive, or it can refer the employee to some reference document, for example company rules and Statutory Wages Orders.

The statement must cover:-

(a) Scale of remuneration;
(b) Hours of work;
(c) Holiday entitlement;
(d) Any terms and conditions relating to sickness, injury, sick pay or pensions;
(e) Minimum periods of notice to be given by the employer (37);
(f) A note explaining any disciplinary rules relating to the employment and the grievance procedures which the employee may use in a dispute;
(g) The job title.

Where a union membership agreement is in existence, it is advisable to refer to this in the statement.

Wherever the employer does not supply a statement within the provisions of the Act, and upon application by an employee, Industrial Tribunals are empowered to determine what particulars ought to have been given, and to incorporate these in the applicant's terms and conditions of employment.

Obviously, in any case before a Tribunal, the question of an employee's 'continuity of employment' may have great significance. The Act provides rules for determining continuity.

Continuity is not broken by a change of employer, provided the business is taken over as a going concern and not merely by disposal of the assets. Continuity is also not broken by re-employment following unfair dismissal or maternity leave, national service (38), or industrial disputes; but in the last-named case any week during any part of which the employee was on strike is deductible.

Certain categories of employees are excluded from the provisions of the Act, for fairly obvious reasons. These are:-

(a) Part-time employees (39) (40);
(b) Registered dock workers;
(c) Share fishermen and merchant seamen;
(d) Temporary employees (41);
(e) Employees working abroad (42);
(f) Police and Armed Forces;
(g) Persons over the age of retirement (40);
(h) Independent contractors working under contracts for service.

---

(37) Under the Employment Protection (Consolidation) Act 1978 these are now:- At least 1 week after 4 weeks' employment; at least 2 weeks after 2 years' employment; and pro rata to 12 weeks (maximum) after 12 years' employment.
(38) National Service Act 1948.
(39) Employees who work for less than 16 hours per week (less than 8 hours per week after 5 years' service) (Employment Protection Act 1975, schedule 16).
(40) Part-time employees and pensioners are however protected against unfair dismissal for an inadmissible reason, despite having no 'contract of employment'.
(41) Employed for less than 12 weeks.
(42) But not employees on offshore installations (Employment Protection (Consolidation) Act 1978).

Redundancy

Redundancy arises when an employer no longer has need of an employee's services. It may arise because there is a shrinking requirement for a certain skill, or for workers producing a service or product which is to be discontinued or run down; or simply because the employer ceases to carry on his business altogether, or at the place where the employee was contracted to work.

An employee claiming an award of Redundancy Payment before an Industrial Tribunal must establish that:-

(a) he had at least 104 weeks' continuity of employment with the employer;
(b) he is not in an excluded category of employment (see above (a) to (h));
(c) he has not unreasonably refused a suitable alternative offer of employment (43);
(d) the reason for his dismissal was redundancy.

Redundancy Payments are related to age, length of service, and payment received for a normal working week (44). They are financed 59% by the employer and 41% by the Redundancy Fund to which all employers contribute with their PAYE returns. If an employer, as frequently happens, should be insolvent, the job security provisions of the Employment Protection (Consolidation) Act 1978 provide that each employee will be paid the remuneration owed to him (45). In addition, the same Act lays down procedures for dealing with redundancies which in essence are designed to prevent the sort of 'press release redundancies' which were such a regrettable feature of the past. Employers planning redundancies are required to consult with the appropriate Trades Unions, in default of which employees can obtain a protective award from a Tribunal. Also, if more than 10 employees are affected, notice must be given to the Secretary of State; there is a sliding scale, depending on numbers affected, as to the length of notice which must be given (46).

12.1.3 The Employment Act 1980 - Trade Union Ballots

(This is a new provision and not merely an amendment to the Employment Protection (Consolidation) Act or the Trade Union and Labour Relations Act).

The Act (47) permits the Certification Officer, under regulations to be made by the Secretary of State for Employment, to provide a contribution from public funds towards the costs which an independent trade union can show it has incurred in conducting a secret ballot for any of the following five purposes:-

(a) to call or end a strike or other industrial action;
(b) to elect the members of an executive committee;
(c) to elect full-time union officials or shop stewards;
(d) to amend its rules;
(e) to reach a decision under the Trade Union (Amalgamations etc.) Act 1964 on amalgamation or transfer.

---

(43) He may accept the alternative employment offered for a trial period of 4 weeks, without losing his entitlement to Redundancy Pay if he considers the alternative to be unsuitable.
(44) One and a half weeks' pay for each year of service over age 41 (up to 65); one week's pay for each year of service between ages 22-40; half a week's pay for each year of service between ages 18-21.
(45) And any Redundancy Payment owed.
(46) 90 days' notice where it is proposed to make 100 or more redundant within 90 days; 30 days' notice where it is proposed to make at least 10 but not more than 99 redundant within 30 days.
(47) Employment Act 1980, section 3.

An employer whose workforce numbers more than 20 employees must, if it is reasonably practical for him to do so, permit a recognised Trade Union to use his premises to conduct such a workplace ballot (48).

12.1.4 The Employment Act 1982

This Act extended and further strengthened some of the provisions in the 1980 Act, in particular balloting, closed shops (UMAs) and trade union immunities.

Confirmation of Closed Shops by Ballot

A provision of the Act which extends further the principle of balloting, is that an employee's dismissal, or an employee's 'action short of dismissal', where the employee refuses to belong to a trade union with which the employer has a Union Membership Agreement, becomes automatically unfair unless the UMA has in the last five years been confirmed by an 80% vote of the workers covered by the UMA or (except in the case of a new UMA negotiated after 15th August 1980), by 85% of those actually voting.

Extension of Individual Rights

The Act further extends individual rights by:-

<u>(1) Detailing fresh situations</u> where it will be unfair to dismiss a non-unionist for refusal to belong to a trade union where a UMA exists.

These now include employees:-
(a) who are non-unionists because of a conflict between a professional code of conduct and possible trade union requirements to take part in industrial action - e.g. solicitors in local government;
(b) who are seeking a remedy under the Employment Act 1980 on grounds of unreasonable exclusion from union membership;
(c) in closed shop situations, where in the five years preceding the dismissal there has not been a secret ballot in which the continuance of the closed shop was supported by 80% of the workers covered by it or 85% of those actually voting in the ballot.

The 1982 Act also makes it clear that the non-unionist cannot be fairly dismissed for refusing to make payments, either to a trade union, or to some other body such as a charity in lieu of union membership.

Similar protection is given to the non-unionist against action taken by the employer to compel union membership, where such action takes a form other than dismissal, e.g. withholding promotion from the non-unionist, etc.

This section effectively restored a right not to belong to a trade union, as a countervailing right to the right to belong.

<u>(2) Increasing the compensation</u> for those unlawfully dismissed from a UMA, and extending to all such cases the remedy of interim relief. This means that in order to bring a claim, the applicant need not have been employed by the employer for any particular minimum qualifying period (49).

---

(48) If an employer does not comply with the request, a recognised trade union may complain to an Industrial Tribunal and, if the complaint is upheld, the Tribunal must make a declaration and may award the union compensation.
(49) This remedy was previously only available to a trade unionist dismissed for his trade union membership or activities.

Remedies for Unfair Dismissal

Type of Award	Circumstance	Is Award Reduceable?	Minimum	Maximum
BASIC AWARD	Unfair Dismissal	Yes, to take account of employee's contributory fault	2 weeks' pay*	30 weeks' pay*
	TU membership or activities(s.59a) Refusal to belong to TU/UMA (s.58)	Ditto	£2,000	ditto
COMPENSATORY	All cases of Unfair Dismissal	Yes, to extent of employee's contribution	------	£7,500
ADDITIONAL COMPENSATION	Non-compliance	No, cannot be reduced on account of applicant's conduct	13 weeks' pay*	26 weeks' pay*
	Non compliance following sex or race discrimination	Ditto	26 weeks' pay*	52 weeks' pay*
SPECIAL AWARD	ss.58/59a (above)	Not reduceable because of applicant's refusal to belong to a TU, or his TU membership and activities, BUT reduceable if he unreasonably prevents re-employment or refuses his employer's voluntary offer of re-employment	£10,000*	£20,000* (based on 104 weeks pay)
	Non-compliance following Unfair Dismissal under ss.58/59a		£15,000	NO UPPER LIMIT (based on 156 weeks pay)

Note:- * - Up to a maximum (January 1983) of £140 per week.

First of all as regards compensation: whereas in ordinary cases of unfair dismissal the minimum basic award is likely to be only two weeks' wages, in the case of the non-unionist the basic award is fixed at £2,000 which, however, may be reduced on account of the employee's own conduct. It is specifically provided, however, that the fact that the non-unionist broke any contractual term, whereby he had agreed to belong to a trade union, is to be ignored when it comes to scaling down any compensation awarded to a non-unionist. Compensation is of course not confined to the basic award; the non-unionist will be entitled as before to claim the normal compensatory award.

On top of this compensatory award there is a new 'special award' for the non-unionist. This award will be made where the dismissed non-unionist asks the tribunal to order his reinstatement or re-engagement but no reinstatement or re-engagement has occurred. Where this is because the tribunal decided not to order the employer to give the non-unionist his job back, the amount of compensation awarded will be 104 weeks' wages or £10,000, whichever is greater, subject to a maximum of £20,000. Where the tribunal has ordered the employer to give the non-unionist his job back, and the employer has without good excuse failed to comply, then the amount of the special award is to be 156 weeks' wages or £15,000, whichever is greater. In this case there is no upper limit to the award that can be made. The special award can be scaled down to take account of the dismissed worker's own conduct.

(3) 'Joining' of Trade Union The non-unionist in the closed shop who has been dismissed or discriminated against on account of his non-membership may not only bring a claim for compensation against the employer concerned, but may also 'join' as a defendant any trade union, officer or member of a trade union, who has put pressure by threatening industrial action against the employer, to persuade the employer to take action against the non-unionist.

Instead of ordering the employer to pay the non-unionist compensation, the tribunal can order the joined defendant, whether trade union or trade unionist, to pay all or part of the compensation to the non-unionist.

(4) Retrospective Compensation for those unfairly dismissed from closed shops in the past.

The Act allowed any person who was dismissed on account of the existence of a closed shop in the period 16th September 1974 to 14th August 1980 inclusive, at a time when the then legislation made such dismissals fair, to bring a claim for compensation in respect of such dismissal, on condition that the changes introduced in the law governing the dismissal of non-unionists in closed shop situations by the Employment Act 1980 would, if they had been in force, have rendered such a dismissal unfair. Workers had 12 months from October 28th 1982 within which to make written applications for compensation to the Secretary of State for Employment.

Trade Union/Employer Relationships

The Act also further regulates this relationship by:-

(1) Reducing still further the immunities available to trade unions when they take action in furtherance of a trade dispute, and bringing these immunities into line with those available to individual trade unionists.

The equivalent of a right to strike is established in British law by conferring immunities upon trade unions and trade unionists who organise industrial action. As regards trade unions, this has taken the form of giving them a complete immunity from all claims for damages for breaches of contracts of employment and commercial contracts (e.g. by a 'sit-in'). These latter were brought within the scope of the immunities by section 13(2) of the Trade Union and Labour Relations Act 1974, but this section is specifically repealed by the Employment Act 1982, thus making it easier for an employer to seek injunctions against trade unions and their organisers.

This immunity for trade unions has now been abolished, and trade unions can be sued in the same way as trade unionists, but the Employment Act 1982 sets upper limits to the amounts of damages that may be awarded against a trade union in legal proceedings. The upper limit is £10,000 against a trade union with less than 5,000 members; £50,000 against a trade union with between 5,000 and 25,000 members; £125,000 where the union has between 25,000 and 100,000 members; and £250,000 where the union has more than 100.000 members.

Section 17 of the Act protects certain union private property and funds, such as provident funds, from sequestration (being seized to pay) for damages. The Act provides that a trade union is to be held liable for any unlawful action authorised or endorsed by its Executive Committee, President, General Secretary or any of its officers with authority to call industrial action under the union's rules. A trade union is also liable for any unlawful action authorised or endorsed by any of its employees, or any committee to which such employed officials report, except where the authorisation is promptly over-ruled by higher union authority (i.e. Executive Committee, President or General Secretary, provided the 'repudiation' takes place as soon as practical and the relevant committee/official is given written notice of it), or where the union rules prohibit the employed officials or committee from calling industrial action.

The 'right to strike' depends crucially on how the law interprets the term 'trade dispute'. The 1982 Act now restricts the scope of the term 'trade dispute' by bringing it specifically within the definition to be found in the Trade Union and Labour Relations Act 1974, which confines (50) the subject matter of trade disputes to a specific list of topics. These include such things as:

(a) Terms and conditions of employment;
(b) The physical working conditions;
(c) Facilities for shop stewards;
(d) Discipline, dismissal, and suspension of workers;
(e) Allocation of work;
(f) Machinery for negotiation and consultation.

Specifically excluded from the notion of a 'trade dispute' are disputes to compel employers to employ union labour only on the performance of commercial contracts, and disputes relating to matters occurring outside the UK, unless the persons taking industrial action are likely to be affected by the outcome of the dispute.

(2) Selective dismissal of strikers

In the past, workers dismissed for taking part in industrial action could bring claims for unfair dismissal, if the employer had discriminated in dismissing some but not all strikers, or had discriminated in selectively offering re-employment to only some of those who had been dismissed.

The 1982 Act enables an employer to be selective as to which dismissed strikers are re-engaged, by issuing an ultimatum to workers taking industrial action. Provided he dismisses all who are still taking industrial action after the ultimatum has taken effect, then none of those dismissed can bring a claim for unfair dismissal, provided also that the employer has not, within three months of the dismissal, selectively offered re-employment to some of those dismissed.

A complaint that an employer has re-engaged another 'relevant' striking employee within the 3 months must be made within 6 months of the complainant's dismissal, but the complainant cannot now compare himself to a re-engaged striker who:

(a) was working at a different establishment, or
(b) was working at the same establishment, but not striking or taking other industrial action at the time of the complainant's dismissal.

(3) Banning union-only labour contracts

The Act makes void any commercial contract term which requires an employer to use only union (or only non-union) labour in performing the contract. Similarly, it makes it unlawful to exclude from a tender list (or even to refuse to offer a contract) on the grounds that people employed on fulfilling the contract are or are not trade union members.

---

(50) Section 29.

It also makes void any provision in a commercial contract to require any employer to have dealings with any trade union. It is unlawful to refuse to contract with any person because of their refusal to deal with trade unions.

(4) Statements on Employee Involvement

The Act requires the Directors of Companies with more than 250 employees in the UK to include in their Annual Reports a statement of action taken to introduce, maintain or extend employee involvement.

The statement must show what steps have been taken in relation to the:-
(a) systematic provision of information to employees;
(b) regular consultation with employees or their representatives 'so that the views of employees can be taken into account in making decisions which are likely to affect their interests';
(c) encouragement of employee involvement through share schemes, etc;
(d) achievement of employee awareness of the economic factors affecting the company's performance.

This requirement is a reflection and a reaction to the commitment of the EEC to the introduction of minimum standards of worker participation in member countries.

12.1.5 Codes of Practice

The Industrial Relations Code of Practice issued in 1972 by the Department of Employment continues to have effect under the Trade Union and Labour Relations Act (51) until superseded by Codes on specific topics prepared by ACAS (52) or by the Secretary of State (53). Whilst it is not in itself either a criminal or a civil offence to fail to observe the provisions of a Code, courts and Industrial Tribunals will be able to take into account any such failure in any proceedings in which the Codes have relevance (54).

The Equal Opportunities Commission, Commission for Racial Equality and Health and Safety Commission (see below for all three of these) all have similar Code-making powers.

12.2 Industrial Training

The Industrial Training Act 1964, which initially created Industrial Tribunals, had three main aims:-
(i) to improve the quality of training;
(ii) to provide an adequate supply of trained manpower;
(iii) to spread the cost of training.

To this end 27 Industrial Training Boards were created. Financed by means of a statutory levy on all employers, they make grants in respect of training carried out. While the system obviously went far towards meeting the aims of the Act, serious imperfections became apparent and in 1973 there was further legislation,

---

(51) Also Employment Protection (Consolidation) Act 1978, schedule 17; and Employment Act 1980, section 3.
(52) Employment Protection (Consolidation) Act 1978, section 6. The Secretary of State and Parliament must both approve the draft Code prepared by ACAS after consultation before it is issued. See also footnote 14.
(53) Employment Act 1980, section 3. The Secretary of State must consult with ACAS before proceeding to issue a draft Code. See also footnote 14.
(54) For example, the Secretary of State's draft Code on Picketing suggests a maximum of six pickets is necessary at a factory entrance.

the Employment and Training Act. This set up a Manpower Commission to which reported an Employment Services Division (55) and a Training Services Division (55). In general Industrial Training Boards were encouraged to disengage from the levy/ grant mechanism by, for example, placing ceilings on the amount of levy they could raise, exempting small employers from levy, and allowing large employers who satisfy their parent Board's training criteria to opt out of the system. The Training Services Division can make specific grants to encourage certain types of training (e.g. the training of young HGV drivers). In addition employers may combine together with Group Training Associations, of which there are many in the road transport industry, to co-operate for the purposes of training, and the Training Services Division itself runs many Government 'Skill Centres'. The Training Opportunities Scheme (TOPS) is available for retraining unemployed persons either through Skill Centres, Group Training Associations or educational establishments.

A Special Programmes Division was created within the MSC in the late 1970s, charged with assisting jobless school leavers and unemployed adults.

The Employment and Training Act 1981 and the Industrial Training Act 1982 were instrumental in permitting the Secretary of State for Employment to commission from the MSC a sector-by-sector appraisal of Industrial Training and, acting on this, to wind up individual ITBs. The Air Transport and Travel ITB has been wound up in this way, and the Road Transport ITB has had Road Passenger Transport, Removals and Storage, and Driving Schools removed from its scope, leaving a slimmed-down Board for the Haulage and Garage sectors of the industry.

A New Training Initiative (NTI)

The MSC's consultative document (May 1981) suggested three objectives of future policy:-

(i) the development of skills training, including apprenticeships, with entry at different education levels and ages, and enabling young people and adults to acquire agreed and appropriate standards;

(ii) a guarantee to all 16-18-year-olds of either full-time continuing education or training, or a period of planned work experience with related education and training;

(iii) the creation of widespread opportunities for adults to update their knowledge and skills during their working life.

Arising out of the consultative document is the establishment of the YTS and the Open Tech. The first objective is more difficult, having considerable implications in the field of industrial relations, especially for craft unions.

12.3 Discrimination in Employment

Legislation (56) enacted during the last two decades has the effect of making it unlawful to discriminate on the grounds of race or sex in such areas as employment, trade union membership, education and training, housing, and the provision of services and facilities (57). Individual allegations of discrimination in employment (58) are dealt with by Industrial Tribunals.

(55) All of these obtained Crown status under the Employment Protection (Consolidation) Act 1978.

(56) Race Relations Acts 1965 and 1976; Sex Discrimination Act 1975.

(57) E.g. supply of credit, membership of clubs (except clubs whose essential purpose is to be a group of individuals of a particular race), education.

(58) Recruitment, terms of contract, facilities, promotion, training, dismissal. Includes also complaints to an Industrial Tribunal under the Equal Pay Act 1970 (The County Court handles complaints in the fields other than employment).

12.3.1 The Equal Pay Act 1970

The Act requires that men and women who perform the same or 'broadly similar' (59) work of equal value (60) should receive equal treatment as regards their terms and conditions of employment (61).

12.3.2 Sex Discrimination Act 1975

The Act makes discrimination on the grounds of sex or marital status unlawful. An Equal Opportunities Commission (EOC) is established to help enforce the legislation and promote equality of opportunity. The Commission may assist individuals, conduct formal investigations (62) and deal with discriminatory practices as appropriate, either by formal investigation, reference to Industrial Tribunals, or prosecution.

12.3.3 Race Relations Act 1976

The Act defines unlawful discrimination as treating a person less favourably on the grounds of colour, race, ethnic or national origins. A Commission for Racial Equality (CRE) (63) is established with similar powers to the Equal Opportunities Commission.

12.3.4 Discriminatory Practices

Only the Equal Opportunities Commission and the Commission for Racial Equality (64) may bring legal proceedings for conduct which, while not in itself unlawful, is designed to produce, or might result in, unlawful discrimination against a race etc. or sex. Both the Sex Discrimination and Race Relations Acts make employers liable for the discriminatory acts of their employees, unless they can show in defence that they took all practical reasonable steps to prevent such acts.

12.3.5 Indirect discrimination

This is treatment which, while equal in a formal sense, is discriminatory in its effect upon the individual on the grounds of race etc. or sex, or upon a particular race etc. or sex as a whole. It arises where a person applies an unreasonable condition which is such that the proportion of persons of the race or sex against which discrimination is alleged who can comply, is considerably less than the proportion of persons not of that group who can comply.

---

(59) Or different work given equal value under a job evaluation scheme. Comparisons may now be made with the terms and conditions of a previous employee in the same post (Smith v. McCathy's (Wembley) Ltd, European Court of Justice 1980).

(60) Article 119 of the Treaty of Rome. In McCathy v. Smith, 1980, it was held that a woman may make comparisons with her predecessor's pay.

(61) Unless different pay and conditions arise because of
- (a) Maternity provisions of the Employment Protection Act, or Regulations on hours of work or shifts (but employers may apply for special exemption certificates for female employees to work night shifts);
- (b) Differences in retirement ages affecting pension and death benefits;
- (c) Marriage benefits for women which were negotiated before 12.11.75.

(62) And if necessary issue non-discrimination notices.

(63) Replacing the Race Relations Board and Community Relations Commission of the 1965 Act.

(64) Both the Equal Opportunities Commission and the Commission for Racial Equality have additional powers to produce relevant Codes of Practice.

This should not be confused with positive action, which both the Sex Discrimination and Race Relations Acts permit; this amounts to providing access to, and encouragement to use, training facilities for a particular racial group or minority sex, where in the previous 12 months the proportion of persons in that group in relation to the workforce as a whole was small (or indeed where no such persons were employed in the period) (65).

It is unlawful to advertise for employees in any way which indicates an intention to discriminate, or which uses a job description with sexual connotations.

12.3.6 Exemptions

There are exemptions under both Acts to the requirements that employers must not discriminate against their employees or potential employees:-
(a) Domestic employment in private households;
(b) Certain welfare services intended for a particular racial group or sex which can be provided most effectively by persons of the same race or sex;
(c) Where a person's sex or race is a Genuine Occupational Qualification because of, where applicable (66):-
  (i) the essential nature of the job (67) or
  (ii) considerations of authenticity, accommodation (e.g. single-sex or shared), decency, privacy, national security (67);
(d) Employment outside Great Britain.

Under the Sex Discrimination Act also:-
(e) Employment in small firms with under 5 employees.
(f) Certain discrimination arising naturally from a woman's pregnancy or retirement.

Under the Race Relations Act also:-
(g) Recruitment of seamen abroad.

12.3.7 Individual enforcement

An Industrial Tribunal deciding in favour of a complainant under either Act may award the following remedies:-
(a) An order declaring the rights of the parties;
(b) A recommendation that the respondent takes a particular course of action;
(c) An award of compensation (68) (69).

There is a provision in both Acts involving the use of statutory questionnaires and reply forms which claimants can send to the respondent for his reply when they consider they may have been the victims of discrimination. The intention is to help them to decide whether to institute proceedings, by focussing attention on the relevant facts.

---

(65) Replacing the 'quota' provisions of the 1965 Race Relations Act.
(66) Strength or stamina is not a Genuine Occupational Qualification.
(67) Examples of jobs where sex or race are Genuine Occupational Qualifications include:- Model, Toilet Attendant, Actor, Waiter in an ethnic restaurant, Civil Servant in Ministry of Defence, Miner.
(68) Composed of a basic award (based on the Redundancy Payments scale maximum 30 weeks at (January 1983) £140 per week = £4200), plus a compensatory award (maximum £7280 for 52 weeks), and up to the same amount again if the employer fails to comply with a reinstatement or re-engagement order.
(69) Including compensation for hurt feelings, but not exceeding the maximum compensation payable by Tribunals. An additional award of 13-26 weeks' pay is made if the act complained of constitutes both unlawful discrimination and unfair dismissal (see Figure 27).

12.3.8 The Disabled Persons (Employment) Acts 1944 and 1958

Employers with more than 20 employees have an obligation to employ not less than 3% of Registered Disabled Persons unless, because of the nature of the jobs, an exemption certificate is obtained from the Secretary of State for Employment.

Certain jobs are designated to be filled only by Registered Disabled Persons unless the employer obtains a permit to employ other persons.

12.3.9 Rehabilitation of Offenders Act 1974

The Act allows convictions for criminal offences involving a prison sentence of not more than two and a half years, to become 'spent' after a specified period of time during which the offender has not again been convicted. An employer may not discriminate against an employee or potential employee on the grounds of a 'spent' conviction.

12.4 Social Security Legislation

National Insurance is a development of the system resulting from the Beveridge Report of 1942, which combined State insurance against injury at work with State welfare insurance (70).

Contributions to the scheme are made by the employer and employee through the PAYE system (71). The State Graduated Pension Scheme introduced in 1961 was wound up in 1975, to be replaced by April 1978 by a new State Pension Scheme (72).

The main benefits covered by National Insurance are:-
(a) Unemployment and Sickness Benefits;
(b) State Pensions;
(c) Industrial Injury Benefits;
(d) National Health Service;
(e) Redundancy Payments.

12.4.1 The Social Security Act 1975

The Act sets out the conditions under which benefit is payable. With some exceptions (e.g. unemployment, sickness and industrial injury) an employee is not 'in benefit' unless a minimum number of contributions have been made, although an employee 'out of benefit' may claim Supplementary Benefit for himself. (73) and his dependents. Disqualification from benefit can arise in a number of ways:-
(a) In the event of unemployment as a result of a trade dispute, unless the claimant can show that he was not participating in, nor financing, the dispute (74).
(b) If the employee is unemployed because of industrial misconduct; in this case he will be disqualified from benefit for 6 weeks.
(c) If the employee refuses to accept suitable employment offered, provided the vacancy has not arisen as a result of a trade dispute or because the rate of

---

(70) Social Security Act 1975.
(71) Previously only Graduated Pension contributions were collected in this way; flat-rate contributions were made by purchasing National Insurance stamps and sticking these to cards.
(72) Social Security Pensions Act, 1975.
(73) Subject to his not actually being disqualified; even then, in principle, his dependents will not be deprived of benefit. However, if the claimant is unemployed because of a Trade Dispute he will be 'deemed' to be receiving strike pay from his Union (Finance Act, 1979).
(74) Social Security Act 1975, section 19.

pay is less favourable than the recognised terms (75).

The Acts also provide that workers supplied by Employment Agencies are treated as employed persons, the Agency being liable for their National Insurance contributions. Employers are responsible for paying combined contributions for persons not in their direct employment (76), even though they may be in the service of an intermediate employer.

Industrial Injury All persons in insurable employment are insured against personal injury caused by an accident arising out of and in the course of their employment, or by prescribed diseases. Whilst Industrial Injury Benefit was abolished in 1982 (77), so that employees injured at work must now rely on ordinary sickness benefit, Industrial Disablement Benefit remains payable to those suffering longer-term disability as a result of an industrial injury. Disablement benefit is not dependent on prior payment of National Insurance contributions.

To be 'in benefit' the accident must have arisen out of the employment (78), i.e. while doing something the claimant is employed to do, and in the course of employment, i.e. after employment commences and before its termination. Travelling to work in the employer's vehicle and with his permission is deemed to be 'in the course of employment' (79). A claimant injured whilst acting in an emergency will also be 'in benefit'.

Administration Since the National Insurance Act 1965, the administrations of the National Insurance Act and National Insurance (Industrial Injuries) Act (79) have been integrated. Initially the Local Insurance Officer either allows the claim, disallows it (in writing), or refers it to a Local Appeals Tribunal (80). This consists of an independent Chairman, a lay member with 'knowledge of local conditions' (usually from the CBI), and an employee (usually trade unionist) member. From here, appeals by either the claimant or the Insurance Officer go to the National Insurance Commissioners, who may deal with them individually or call a tribunal of Commissioners. In medical cases a local Medical Board replaces the National Insurance Tribunal. On points of law, appeals from decisions of the Commissioners are dealt with by the High Court.

12.4.2 Pensions

In April 1978 there was introduced a two-tier system of State Pensions, consisting of a flat-rate basic level pension and an additional 'earnings related' pension. The latter is built up between 1978 and retirement and linked to all earnings in what is known as the 'Top Band'.

The 'Lower Band' (lower earnings limit) is related to the flat-rate pension which is thus equal to 100% of earnings up to this basic level. The upper earnings limit (above which earnings are not pensionable) is approximately seven times this basic level. The limits are fixed anew at the start of each tax year, with both the basic pension and the earnings-related pension being inflation-proofed

---

(75) As with unemployment as a result of industrial misconduct - 6 weeks' disqualification.
(76) i.e. if they are under the employer's general control and management - 'lump' workers.
(77) Social Security and Housing Benefits Act 1982, section 39.
(78) Regina v. Industrial Injury Commissioners (ex parte AEU), 1966.
(79) National Insurance (Industrial Injuries) Act 1965, section 9; now replaced by the Social Security Act 1975.
(80) Now merged with the Social Security Appeals Tribunal, which is independent of the DHSS (Social Security Adjudication Act, 1983).

in line with the Retail Prices Index.

Employers are able to contract out of part of the new arrangements by substituting benefits under a company plan (if approved by an Occupational Pensions Board) for the 'Top Band' (earnings-related) pension. Contracted-out employees, and their employers, pay a lower rate of National Insurance contributions on earnings in the 'Top Band'.

12.4.3 Self-Certification of Sickness and Statutory Sick Pay (77)

From 1st June 1982, Doctors' Certificates have not been required for absences of less than 8 days. Instead, employees must complete self-certifications. After 7 days a Doctor's note is required. No benefit is paid for the first 3 ('waiting') days, unless such waiting days have already occurred in a person's previous 'linked' absence, i.e. where the employee has been at work for no more than 14 days between the two periods of sickness. From 6th April 1983, employers have been responsible for paying Statutory Sickness Pay (SSP) during the first 8 weeks of sickness absence by any employee in any one tax year; they may claim this back through deductions from their National Insurance payments. After 8 weeks sickness have been paid for, the employee is 'transferred' and, if still sick, reverts to claiming Sickness Benefit from the DHSS as under the previous system.

12.5 Safety in Employment

In 1972 the Robens Committee (81) on Health and Safety at Work reported to the Government. Existing legislation (82), the Committee argued, was overcomplicated, did not cover all employees, and quickly became outdated. As a result of this report, the Government passed:-

12.5.1 The Health and Safety at Work etc. Act 1974

The Act applies to all work premises and activities, and affects all employers, every person in control of work premises, and designers, manufacturers, importers and suppliers of equipment designed for use at work, and all employees.

Employers have a duty to provide a safe system of work. This includes a duty on the employer to conduct his undertaking in such a way that as far as is reasonably practicable, persons not in his employment who may be affected are not exposed to risks - a provision which should not be ignored by the road freight transport operator. Employees have a duty to co-operate with the employer as he complies with this duty.

The Act sets up a Health and Safety Commission to act as advisers and issue Codes of Practice etc., and a Health and Safety Executive responsible for enforcing the Act and regulations. There is provision for the establishment of Safety Committees where Safety Representatives request this (83), and for employers to prepare for the notice of their employees a written statement of their Safety Policy (84).

The Health and Safety Executive is given powers of enforcement including:-
(a) the issue of improvement notices;
(b) the issue of prohibition notices;
(c) the power to 'seize and render harmless' any dangerous article or substance;

---

(77) Social Security and Housing Benefits Act 1982, section 39.
(81) Cmnd. 5034.
(82) Factories Act 1961; Offices, Shops and Railway Premises Act 1963, section 9.
(83) Health and Safety at Work etc. Act 1974, section 2(7).
(84) Health and Safety at Work etc. Act 1974, section 2(3).

(d) prosecution.
Appeals against (a) and (b) above are heard by Industrial Tribunals.

12.5.2 Notification of Accidents and Dangerous Occurrences

The Notification of Accidents and Dangerous Occurrences Regulations 1980 place an obligation on employers to keep records, and to provide the DHSS with information relating to dangerous occurrences and accidents. In particular, employers must:-

(1) Provide and keep available an <u>Accident Book</u> (BI 510) for the use of employees and their representatives.
(2) Separately record minimum statutory details of all <u>notifiable accidents</u> (85) to employees, sub-contractors, visitors and the public, and preserve this record for at least three years.
(3) Similarly record all <u>notifiable dangerous occurrences</u> (86), and preserve this record for at least three years.
(4) Record in a <u>General Register</u> particulars of all notifiable diseases contracted by employees, visitors etc., and preserve this record for 2 years.
(5) Record all enquiries from the DHSS about the above occurrences, and past employees, and keep this record for at least three years.
(6) <u>Immediately</u> contact the Health and Safety Executive to notify them of all reportable accidents, and within 7 days confirm full details to them on Form F2508.
(7) If the activities carried out at an operator's premises go beyond repair and maintenance - for example to include full vehicle overhauls - the premises should be registered (using General Register form 31) as a factory with the Health and Safety Executive.
(8) Other useful forms are:-
Form 1 (Abstract of Factories Act);
Form OSR9 (Abstract of Offices, Shops and Railway Premises Act);
Form 954 (Electricity Regulations) (87);
Form 11 (Hours of Work - Women and Young Persons).

12.5.3 Codes of Practice

A fundamental 'prop' of the Health and Safety at Work etc. Act is the concept of Codes of Practice. When the Commission adopts a particular Code of Practice (88), non-compliance with its provisions, while not in itself an offence, can be taken into account in determining guilt or innocence in any prosecution brought under the Act.

The advantages of using Codes of Practice are that
(a) they are current and relevant; and
(b) their implementation is not wasteful of Parliament's time, and they can therefore be of more immediate impact than Statutory Instruments.

That there is a need for this legislation cannot be doubted. In an average year there are approximately 20 million working days lost due to industrial injuries (89).

---

(85) i.e. death or major injury (including most fractures), or any accident resulting in more than 3 consecutive days' incapacity for work.
(86) Includes any collapse of a lift, hoist or crane, or any fire, explosion or electrical short-circuit causing a shutdown of more than 24 hours.
(87) Plus Electric Shock Treatment Posters, available from insurance companies.
(88) Examples within the Road Transport Industry might be the RHA/FTA Code of Safe Loading, and the HAZCHEM Code on carriage of dangerous liquids.
(89) Compare days lost by industrial disputes 1975-1983: 10 million days per annum (average).

13 Information for PSV Operators

General Note

This chapter gives details of legal and other considerations applying to PSV operation wherever this information varies from that for Goods Vehicle operation, detailed in chapters 1-12 of this book. The section numbers in this chapter indicate the numbers of the corresponding sections in chapters 1-12, thus section 13.1.2.2 is the PSV equivalent of section 1.2.2 in Chapter 1 relating to Goods Vehicles. Where there is no equivalent section in this chapter, the material in the earlier section of the book may be taken as referring to PSVs as well unless the context plainly forbids.

13.1 PSV Operator Licensing and Road Service Licensing

13.1.2 PSV O-Licences

The Transport Act 1980 (1) changed the system which had previously existed of licensing individual PSVs, into a system of PSV Operators' Licensing which was modelled very closely on the system of Goods Vehicle Operators' Licensing in the Transport Act 1968. Like its Goods Vehicle counterpart, the PSV O-Licence is a Quality Licence.

13.1.2.1 Scope of PSV O-Licensing

A PSV O-Licence is required to use a vehicle on the road as a stage, express or contract carriage. (See Section 13.1.3, 'Road Service Licensing', below).

A Standard O-Licence, which can be either National or National/International, is required unless the operation can be covered by a:-

Restricted Licence, issued for small passenger vehicles adapted to carry 16 or less passengers, which are used 'non commercially' (2).

13.1.2.2 Applications for PSV O-Licences

These must be made, at least nine weeks before the licence is required, to the Traffic Commissioners for the Traffic Area(s) in which the operator has one or more operating centre(s) (3). A separate licence is required for each Area. Operators may apply to license the maximum number of vehicles which they are

---

(1) Consolidated by the Public Passenger Vehicles Act 1981.
(2) Public Passenger Vehicles Act 1981, section 13. 'Non commercially' in this context means 'not used in the course of business by an operator of PSVs with over 8 seats.'
(3) 'The base or centre at which the vehicle is normally kept.'

likely to need at any one time (4).

Applicants must specify PSVs in possession (5), but may without formality run additional vehicles acquired 'within their margin', i.e. up to the total on the licence, provided that a PSV O-Licence disc is displayed on each vehicle (6).

13.1.2.3 Duration of Licences and Fees

The normal duration of a PSV O-Licence is 5 years (7). A fee of £3.50 per vehicle per month is payable.

13.1.2.4 Matters which the Traffic Commissioners are required to consider

The grant of an O-Licence depends on the applicant satisfying the Traffic Commissioners as to his:-
(a) Good Repute (8);
(b) Appropriate Financial Standing (8);
(c) Professional Competence (9).

Applicants are required to make a declaration (10) that they will comply with legal requirements concerning:-
(a) Drivers' hours and records;
(b) Vehicle carrying capacity;
(c) Vehicle mechanical condition;
(d) Having sufficient financial resources to safely operate and maintain the vehicles;
(e) Reporting any relevant convictions of the licence holder, transport manager, drivers, employees or agents.

They are required to furnish the Traffic Commissioners with details of proposed maintenance facilities, or any maintenance contract (11), financial resources, and the means by which the requirements as to professional competence are to be met.

13.1.2.5 Suspension, Curtailment or Revocation of a PSV O-Licence

The Traffic Commissioners may suspend, curtail or revoke an O-Licence on any of the following grounds but, if the holder of the licence requests them to do so, they must hold a public sitting for the purpose. The grounds are that the operator:-
(a) made a false statement to obtain the licence;
(b) has breached any condition on the licence;
(c) his PSVs have been prohibited because of defects;
(d) his PSVs have been used under such a prohibition;
(e) his material circumstances have changed;
(f) he is no longer of good repute or of appropriate financial standing.

---

(4) If this number is greater than the number 'in possession', they are said to have a licensed 'margin'.
(5) Using Form PSV 421A.
(6) Vehicles hired temporarily (for up to 14 days) with or without driver, may be run 'On Hire' using the disc of the operator from whom they are hired - but that operator then remains the licensed 'user' of the vehicle.
(7) This can be reduced, at the Traffic Commissioners' discretion, to suit their administrative convenience.
(8) The criteria are similar to those for Goods Vehicle O-Licensing (q.v.).
(9) Not required for the grant of a Restricted Licence. See also Chapter 2.
(10) By answering questions on the Application Form PSV 421.
(11) The operator does not thereby cease to be the responsible 'user' of the PSV. (Public Passenger Vehicles Act 1981, section 5).

Operators must always (12) notify the Traffic Commissioners of any change of circumstances, including any relevant convictions of the licence holder or his employees, any accident or damage to a PSV likely to affect passenger safety, and any notifiable alterations to a PSV, or to the composition of his fleet.

13.1.2.6 Conditions applicable to Licences

Traffic Commissioners have wide discretion to attach conditions to an O-Licence, including the maximum number of vehicles to be specified, and the places at which they may pick up and set down passengers (13).

13.1.3 Road Service Licences

13.1.3.1 Definition of a PSV

A Public Service Vehicle is a Motor Vehicle (14) which is either

(a) adapted to carry 9 or more passengers and used to carry passengers for hire and reward (15) or

(b) adapted to carry 8 or less passengers and carrying passengers for hire and reward (15) at separate fares in the course of a business of carrying passengers.

13.1.3.2 Car Sharing

A journey is not treated as being in the course of a business of carrying passengers (see (b) above) if the fare(s) paid total less than the vehicle's running costs (16). Neither is a vehicle with 8 or fewer passenger seats treated as a PSV simply because passengers are carried at separate fares, provided that (17):-

(a) the driver, owner or provider of the vehicle has not made the arrangements to pay separate fares;

(b) there has been no previous advertisement of the journey (18), except where a Local Authority approves the operation as necessary to meet the social and welfare needs of the community;

(c) the journey is not an extension of one made under a Road Service Licence.

13.1.3.3 Contract Carriage Operations

No Road Service Licence is required to operate a PSV as a contract carriage (19), i.e. where no separate fares are charged.

However a journey, even a regular service, may be operated at separate fares and

---

(12) Including any time between application and grant of an O-Licence. (Public Passenger Vehicles Act 1981, section 20).

(13) Public Passenger Vehicles Act 1981, section 16. It has been suggested that this power might be used to regulate the operation of Express Services.

(14) Excluding a tramcar (Public Passenger Vehicles Act 1981, section 1).

(15) 'Hire and reward' includes any payment for carriage made directly or indirectly via any agency, or other matter additional to but including carriage, even if the right to be carried is not exercised.

(16) The running costs include depreciation. The vehicle would not then be regarded as a PSV.

(17) Public Passenger Vehicles Act 1981, schedule 1, Part I.

(18) Public Passenger Vehicles Act 1981, schedule 1, Part IV. The ban on advertisements does not include a notice at a place of worship, a place of work, or a club or society; or in a journal circulating mainly among those attending the above places.

(19) However, a PSV O-Licence is necessary.

FIG. 29 DEFINITION OF A P.S.V.

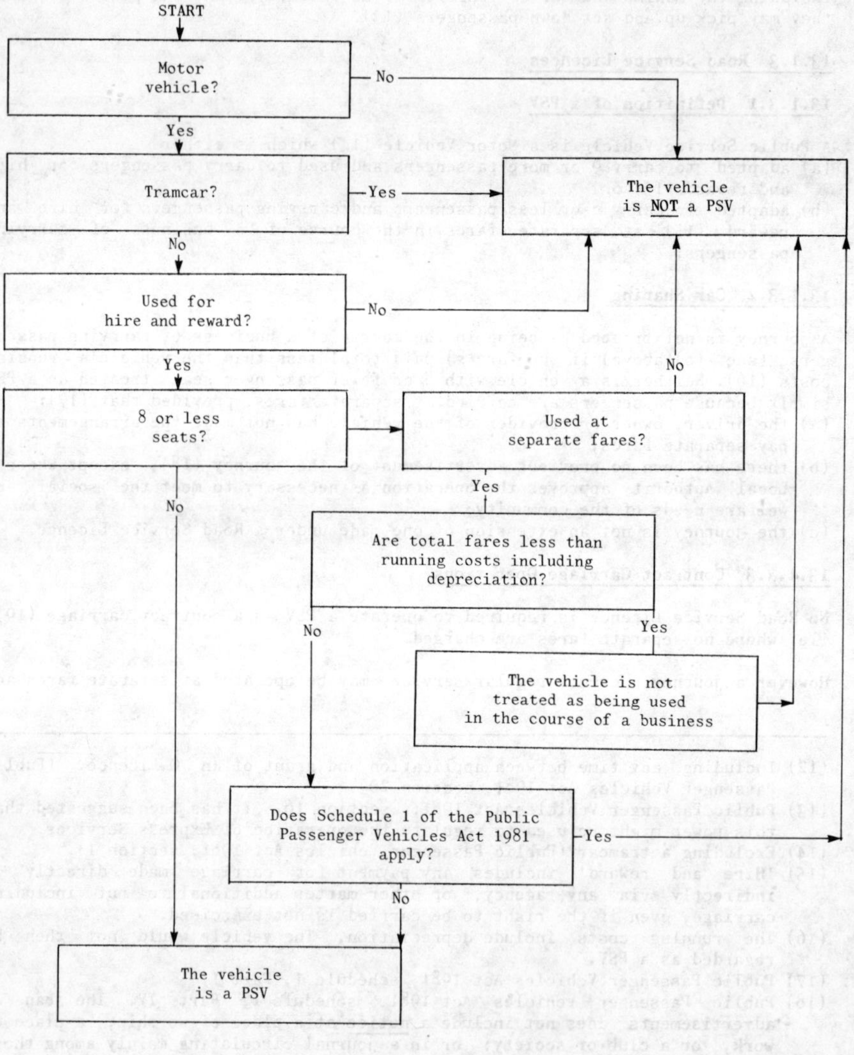

still be regarded as a Contract Carriage provided that (20)

(a) the driver, owner, provider or user (i.e. the O-Licence holder) did not make the arrangements for bringing the party together; **and**
(b) there had been no prior advertisement (18); **and**
(c) there was no differentiation of fares on the basis of distance travelled; **and**
(d) all passengers were carried to the same destination or vicinity;

or

(e) the passengers are a party of overseas visitors, for whom the arrangements to carry them were made before their arrival in Great Britain.

13.1.3.4 Express Carriage Operations

No Road Service Licence is required to operate an Express Carriage Service (21), although the Traffic Commissioners do require notification of certain particulars before the service is commenced or terminated (22). An Express Carriage is a PSV used on an Express Service, i.e. where

(a) no passenger is set down within 30 miles (as the crow flies) of where he is picked up;
(b) the route takes the vehicle more than 30 miles (as the crow flies) between two points (23).

13.1.3.5 Stage Carriage Operations

A stage carriage is a PSV used on a **local service** (i.e. where passengers are carried at separate fares and the service is NOT an Express Service). A **Road Service Licence** is required for a Stage Carriage Service.

13.1.3.6 Applications for Road Service Licences

Applications should be made to the Traffic Commissioners at least 12 weeks before it is intended to operate the service (24). The application is published by the Traffic Commissioners in 'Notices and Proceedings' (25).

13.1.3.7 Grant of a Road Service Licence

Traffic Commissioners must grant a Road Service Licence if

(a) they are satisfied that there are no other transport services or facilities to serve the route (26) **or**
(b) the service is an Excursion & Tour not in competition with any other bus or

---

(18) Public Passenger Vehicles Act 1981, schedule 1, Part IV. The ban on advertisements does not include a **notice** at a place of worship, a place of work, or a club or society; or in a journal circulating mainly among those attending the above places.
(20) Public Passenger Vehicles Act 1981, schedule 1, parts II and III. (These conditions also apply to the operator of a passenger vehicle with 8 or less seats, operated at separate fares so as not to be a PSV).
(21) However, a PSV O-Licence is necessary. If the Express Carriage is an Excursion & Tour it is unnecessary to notify the Traffic Commissioners of it.
(22) Description of route; timetable; stopping places; whether partly 'stage carriage' or international.
(23) This covers an Excursion & Tour arriving back at the departure point.
(24) Using Form PSV8 (PSV9 if an Excursion & Tour, PSV89 for variations to an existing Road Service Licence).
(25) Fortnightly.
(26) The Road Service Licence must contain a statement to this effect, whereupon no appeal is allowed.

FIG. 30 TYPES OF P.S.V.s

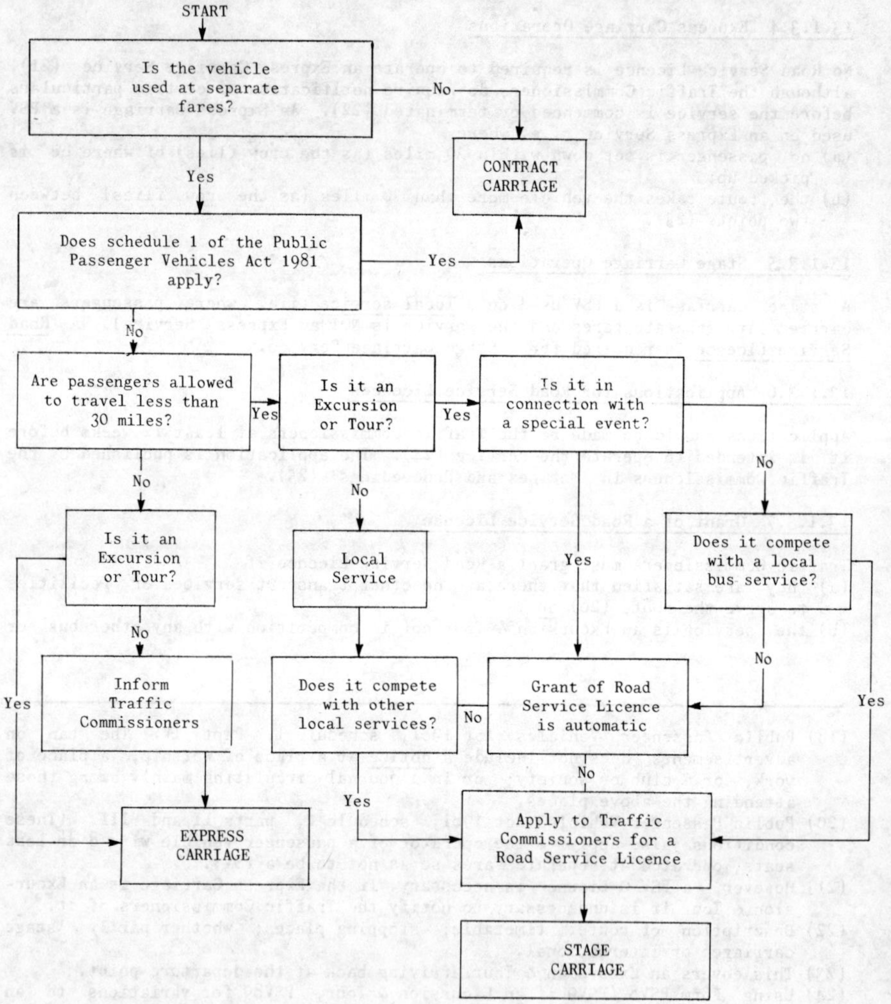

tramcar service (27) or
(c) it is operated only in connection with a special event.

Traffic Commissioners must grant a Road Service Licence unless they are of the opinion that to do so would not be in the public interest (28), having regard to
(a) the transport requirements of the area;
(b) any objections or representations made to them;
(c) Transport Policies and Plans drawn to their attention by Local Authorities.
They must advise County and District Councils, and the Police, of every licence which they grant (29), suspend or revoke.

13.1.3.8 Conditions attached to a Road Service Licence

Traffic Commissioners may attach any conditions which they think fit, especially to ensure
(a) the suitability of the route;
(b) that copies of timetables and faretables are carried;
(c) that passengers are picked up and set down at specified points;
(d) the safety and convenience of passengers, including the disabled.
These conditions may be varied or dispensed by them at any time if it is in the public interest to do so.

In addition Traffic Commissioners, if they think it is in the public interest, may impose fares conditions, either to regulate the terms of competition, or to protect the public from monopolistic pricing (30).

13.1.3.9 Objections and Representations

These must be made to the Traffic Commissioners not later than 28 days after an application is published, with a copy to the applicant, stating the grounds of objection. Traffic Commissioners may convene a Public Sitting (31) to hear details of the application and any objections or representations (32).

13.1.3.10 Revocation or Suspension

A Road Service Licence may be suspended or revoked by the Traffic Commissioners for any breach of the conditions attached to it (29).

13.1.3.11 Expediency

In special circumstances, such as an unforeseen change in demand, or a diversion of a route, Traffic Commissioners may either
(a) grant a dispensation from the conditions on a Road Service Licence, or
(b) grant a short-period Road Service Licence (33).

---

(27) Except other Excursions & Tours. Thus there can be no objection to a competing Excursion & Tour.
(28) Public Passenger Vehicles Act 1981, section 31. They may grant a partial licence if to do so would be in the 'public interest'.
(29) Public Passenger Vehicles Act 1981, sections 31, 36 and 37.
(30) Generally any fare changes need only to be notified to the Traffic Commissioners 21 days prior to introduction (Public Passenger Vehicles Act 1981, section 33).
(31) Giving at least 14 days notice in 'Notices and Proceedings'.
(32) The right to object is no longer restricted to 'statutory objectors'.
(33) For not more than 6 months, and for special reasons only. The operator must apply 12 weeks before expiry of the licence to continue it as a full Road Service Licence.

13.1.4 Appeals

Appeals against the grant, refusal, revocation or suspension of a PSV O-Licence may be made to the Secretary of State for Transport (34).

Only a County or District Council through whose area a stage carriage service operates, or the provider of transport facilities along or near the route (35), may appeal against the grant of a Road Service Licence (36). Special provisions exist where an appeal is made to the Secretary of State against a refusal of the London Transport Executive to enter into an operating agreement with another operator (37). The applicant must first notify the Greater London Council, the appropriate London Borough, and the Metropolitan Police Commissioner. The Secretary of State may, if he thinks fit, make an order requiring the London Transport Executive to enter into an agreement.

13.1.5 Exemptions from PSV Licensing

In certain circumstances, a vehicle which would otherwise require to be operated either with a PSV O-Licence, or with a Road Service Licence, or both, may be operated without either or both. The types of operation, considered in detail below, are:-

(1) Community Bus Operation;
(2) Permit Operations;
(3) Experimental and Trial Areas;
(4) School buses carrying Fare Paying Passengers.

13.1.5.1 Community Buses

Local bus services may be operated by a body of persons concerned with the social and welfare needs of a community on a non-commercial basis under the following conditions:-

(a) The vehicle used has between 9 and 16 passenger seats;
(b) The vehicle displays a Community Bus 'disc' and meets the safety requirements set out in the Community Bus Regulations 1978; it does not require a Certificate of Initial Fitness, but must have an M.O.T. Certificate if over 1 year old;
(c) No PSV O-Licence is required;
(d) The driver must be over 21 years of age and hold a full (not provisional) driving licence;
(e) The driver must be a volunteer and be unpaid (though he may receive out-of-pocket expenses);
(f) The service is operated under a Road Service Licence.

The consent of the London Transport Executive is not required to operate a Community Bus in London, but the Traffic Commissioners must consult with the Executive before granting a Road Service Licence.

The Commissioners may authorise the operation of Excursions, Tours and Contract or Express work if they are satisfied that this is necessary to provide financial support to the main Community Bus service (38).

---

(34) A similar right exists to appeal against the refusal of a Certifying Officer to issue a Certificate of Initial Fitness (see section 13.8).
(35) Public Passenger Vehicles Act 1981, section 50. The applicant has no right of appeal.
(36) Public Passenger Vehicles Act 1981, section 51. Both the Traffic Commissioners and the applicant may appeal to the High Court <u>on a point of law</u>.
(37) Transport (London) Act 1969.
(38) See 'A Guide to Community Transport', HMSO.

13.1.5.2 Permit Operations

Passenger Vehicles being used non-commercially (39) to carry passengers for hire and reward under a Permit, and by the body specified in the Permit (but not used to carry the public at large), will not be treated as PSVs if they are Small Passenger Vehicles. If they are Large Passenger Vehicles they will be treated as PSVs except in that
(a) No PSV O-Licence is required;
(b) The driver need not hold a PSV driver's licence;
(c) No Road Service Licence is required.

Two classes of vehicle are defined (40):-
(a) small passenger vehicles with more than 8 seats but not more than 16 seats excluding the driver (41);
(b) large passenger vehicles with more than 16 seats.

The Secretary of State may designate bodies appearing to him to be concerned with education, religion or the social welfare of the community to grant permits, and may require such bodies to make returns to him in this respect (42). Traffic Commissioners may also grant permits. Both may attach to permits such conditions as they think fit. Permits remain in force until the designated body ceases to be designated.

Passenger vehicles operated under these provisions must
(a) not carry members of the public at large for profit;
(b) be used under a permit and in accordance with any conditions thereon;
(c) be driven by a driver over 21 years of age, with a full driving licence, and who is 'reasonably experienced';
(d) carry an appropriate permit 'disc' issued under these provisions.

Small Passenger Vehicles must comply with the Minibus (Conditions of Fitness, Equipment and Use) Regulations, 1977. Large Passenger Vehicles must also comply with PSV Fitness Regulations (43).

Permits for Large Passenger Vehicles may only be granted by a Local Authority (44), either to itself or to a body which assists and co-ordinates the activities of community organisations within its area. Permits for Small Passenger Vehicles may be granted by (a) the Traffic Commissioners, (b) a local authority or (c) a designated body (45). Vehicles used under a Permit must carry:-
(a) a driver's notice outlining the conditions of use (46), and
(b) a Permit Disc.

13.1.5.3 Experimental Areas

The Secretary of State may designate as an Experimental Area (within which the requirements of the Act regarding PSVs may be modified) the whole or part of the

---

(39) i.e. not for profit but merely so that revenue covers running costs including depreciation.
(40) Public Passenger Vehicles Act 1981, sections 44-46.
(41) Originally (Minibus Act 1977) these were the only class of Permit Vehicle.
(42) Minibuses (Designated Bodies) Order 1980.
(43) Public Service Vehicles (Conditions of Fitness, Equipment, Use and Certification) Regulations 1981.
(44) A County or District Council, a London Borough Council, or a Scottish Regional Council.
(45) Minibus Permits Regulations, 1977.
(46) ibid., Schedule 3.

area of a Local Authority (here defined as the County Council) (47).

A Designation Order may be effective for a period of not more than two years, but may be further extended for periods of up to two years. If requested by a Local Authority, the Secretary of State must either
(a) designate the area <u>or</u>
(b) extend the order.

Within the Area, the Local Authority may grant authorisations for the use of vehicles for 'hire or reward' (whether or not at separate fares) wholly or partly within the Area. Authorisations may be of two kinds:-

(a) <u>General</u> - applying to private vehicles generally or to private vehicles with a specified number of passenger seats (which must be less than 16 by definition).
(b) <u>Special</u> - applying to specific vehicles, either 'commercial' or 'private', and specific persons.

Schedule 5 (48) sets out the conditions which must be attached to a General Authorisation, and which may be attached to a Special Authorisation.

<u>Commercial Vehicles</u> are vehicles normally used to carry passengers for hire and reward, but having 8 or less passenger seats (49).

<u>Private Vehicles</u> have 16 or less passenger seats, but are not normally (50) used to carry passengers for hire and reward.

A Special Authorisation granted by a Local Authority to cover journeys into areas of other Authorities will have no effect in that area unless approved by the appropriate Local Authority. Vehicles used under either Authorisation are <u>not</u> treated as PSVs for the purpose of the Act.

In deciding whether to grant or revoke a Special Authorisation, the Local Authority must have regard to the fitness and suitability of the vehicles and persons using them.

<u>Revocation of Authorisations</u> A Local Authority may revoke an Authorisation at any time, and must do so if directed by the Secretary of State. However, they must take into account any representations made to them in respect of a Special Authorisation, and the revocation cannot take effect until notice (51) of the fact has been published in a local newspaper. The same condition applies where the Secretary of State makes an order, or a Local Authority passes a resolution, revoking a General Authorisation (52).

13.1.5.4 Trial Areas

The Minister is empowered upon application by a County Council to designate all or part of their area as a Trial Area, for a specified period of not less than 2 nor more than 5 years.

Schedule 4 of the Act (48) relates to the making of applications and for

---

(47) Public Passenger Vehicles Act 1981, sections 47-49.
(48) Public Passenger Vehicles Act 1981.
(49) e.g. Taxis.
(50) Disregarding any use made of the vehicle under the Experimental Area authorisation.
(51) 21 days'notice for a General Authorisation; 56 days' notice for a Special Authorisation.
(52) Public Passenger Vehicles Act 1981, schedule 5.

variations or revocation of designation orders.

There is no requirement to hold a Road Service Licence to operate a Stage Carriage Service entirely within a Trial Area, or for that part which operates partially within a Trial Area. Any conditions attached to a Road Service Licence for the part of the service outside the Trial Area will not apply within it (53).

Before starting to operate a Stage Carriage Service within a Trial Area, or changing or discontinuing an existing service, an operator must inform the affected County and District Councils and also publish details in the local press (54). The duty of operators to co-operate and exchange information (55) will not apply in a Trial Area, but information must be afforded to Passenger Transport Executives and County Councils which is needed to allow them to discharge their function of co-ordinating public passenger transport services (56).

13.1.5.5 School Buses carrying Fare-Paying Passengers

A school bus belonging to a Local Education Authority may be used to carry fare-paying passengers, and as such is not subject to PSV fitness regulations, nor is the driver required to hold a PSV driving licence (57). There are two situations where this may apply (58):-

(a) the carriage of fare-paying passengers on a school journey if room is available (i.e. on a journey on which schoolchildren are carried free of charge);
(b) the operation of a local bus service available to the public at large.

13.3.2.2 PSV Drivers' Licences

It is an offence for the driver of a PSV not to hold, in addition to an ordinary driving licence valid for the vehicle, a PSV Driver's Licence (59). If the PSV is being used in such a way that it is not a stage, express or contract carriage, it is sufficient for the driver to hold an ordinary driving licence (60), but if the vehicle has more than 8 passenger seats, the driver must be over 21 years of age (61). Examples of such 'non-PSV use' would be:- empty running between depots, road testing by fitters, carrying non-fare-paying passengers (e.g. staff buses), or driving a PSV whose use as such has been permanently discontinued (e.g. a preserved vehicle) (62).

The licensing of PSV Drivers is controlled by the Traffic Commissioners of the Traffic Area in which the applicant resides (63).

---

(53) Public Passenger Vehicles Act, 1981, sections 38 and 39.
(54) Ibid., section 40.
(55) Transport Act, 1968, section 24; Transport Act, 1978, section 1.
(56) Public Passenger Vehicles Act, 1981, section 41.
(57) However, a Road Service Licence <u>is</u> required.
(58) Public Passenger Vehicles Act, 1981, section 46.
(59) Ibid., section 22(1).
(60) Road Traffic Act, 1960, section 144, as amended by Transport Act, 1980, section 37 and schedule 5 part I para 4(c); and PSV (Drivers' and Conductors' Licensing) (Amendment) (No. 2) Regulations, 1980.
(61) Road Traffic (Drivers' Ages and Hours of Work) Act, 1976; Road Traffic (Drivers' Ages and Hours of Work) (Amendment) Regulations 1980.
(62) Public Passenger Vehicles Act, 1981, section 1(2).
(63) In the Metropolitan Traffic Area, where previously the Metropolitan Commissioner of Police issued PSV Drivers' Licences, this function is now discharged by the Metropolitan Traffic Commissioners (Metropolitan Traffic Area (Transfer of Functions) Order 1984).

Qualifying for a PSV Driver's Licence - Age Limits A person applying for his first PSV Licence must submit himself for a PSV Driving Test and a medical examination (64). The applicant must have a valid full or provisional driving licence (65), and must satisfy the Traffic Commissioners, on form PSV150, that he is a fit person to drive a PSV. The form calls for details or evidence of his age and previous convictions, other personal particulars, his ability to read and write (66), his knowledge of the Highway Code (67), a medical certificate of his fitness to drive a PSV, a certificate of character signed by two householders who have known him for 3 years, and a certificate from his employer (68).

The minimum age for the driver of a large passenger vehicle (i.e. with seats for more than 8 passengers excluding the driver) is 21 years (61), but such a vehicle may be driven by a driver of 18 years of age where either
(a) he is not carrying passengers, and holds a PSV Licence or is supervised by a PSV Licence holder; or
(b) he is carrying passengers either on a regular service on a route under 50 km in length, or on a National journey using a vehicle with seats for not more than 14 passengers excluding the driver, and he holds a PSV Licence.
The minimum age for the driver of a small passenger vehicle (8 or less seats excluding the driver) is 17 years (61).

The holder of a PSV Driving Licence may drive a rigid goods vehicle not exceeding 10 tons ULW which belongs to a PSV operator and which is being driven on the operator's behalf to aid or recover a broken-down and currently licensed PSV (69).

A licence must be produced on demand to a Police Officer, Certifying Officer or PSV Examiner, or, in default, the driver must produce it within five days as with an ordinary licence, or state where it can be seen within that time.

The licence must be signed by the holder, and his PSV badge must be worn in a conspicuous place when he is on duty on a PSV (70). The Commissioners must be notified within 7 days of any change of address by the licence holder.

The PSV Test may be conducted by a DTp examiner, or an examiner delegated by a PSV operator who has obtained permission to appoint an employee as such, and who employs over 250 drivers. This examiner must be approved by the DTp, and must maintain records of tests conducted. The examiner must be satisfied that the candidate is conversant with the Highway Code, competent to drive the vehicle safely, and able to perform specified movements with it, including stopping the vehicle correctly at a bus stop. The candidate should present himself for the test with a clean and roadworthy vehicle which the examiner accepts as being suitable for the category of licence sought.

---

(61) Road Traffic (Drivers' Ages and Hours of Work) Act, 1976; Road Traffic (Drivers' Ages and Hours of Work) (Amendment) Regulations 1980.
(64) The medical examination is also required (a) if a licence lapses for longer than 12 months; (b) on renewal at 50, 56, 59 and 62 years of age; (c) at any time at the Commissioners' discretion.
(65) In the case of a provisional licence holder, the vehicle must display 'L' plates and the candidate is considered to be taking a 'Dual Test' for his Ordinary and Vocational licences. Failure of the Vocational (PSV) Test precludes his qualifying for an ordinary licence.
(66) For this purpose, the form must be completed in the applicant's own handwriting.
(67) The applicant makes a statutory declaration to this effect on the form.
(68) Or some reason (e.g. owner driver) why this cannot be completed.
(69) HGV (Drivers' Licences) (Amendment) Regulations 1977.
(70) PSV (Drivers' and Conductors' Licences) Regulations 1934.

Classes of PSV Licence The following table is self-explanatory.

Classes 1 and 1A - Any type of PSV.
Classes 2 and 2A - Any type of single-deck PSV.
Classes 3 and 3A - Single-deck PSVs with length not exceeding 27ft 6ins.
Classes 4 and 4A - Single-deck PSVs with length not exceeding 17ft 6ins.
The 'A' suffix denotes a vehicle with automatic or semi-automatic or pre-selective gears (i.e. clutchless). In each case a licence to drive a vehicle without automatic transmission entitles the holder to drive a vehicle in the same class with automatic transmission.

Suspension, revocation, disqualification and appeals

If the Traffic Commissioner issuing the licence considers at any time that the holder is not a fit person, he may suspend or revoke the licence, and both the licence and badge must be surrendered within 5 days (71). If the holder is disqualified from driving, his PSV licence becomes invalid for the period of the disqualification.

Appeals against the above, or against the refusal to grant a PSV Licence, are to a Magistrates' Court (72).

Conductors of PSVs no longer require to be licensed.

13.4 PSV Drivers' Hours and Records

13.4.1 Drivers' Hours

Part VI of the Transport Act 1968 (73), together with regulations made under that Act, contain the main body of legislation relating to the Hours of Work and Duty of drivers of passenger vehicles performing Domestic Journeys and Work. EEC Regulation 543/1969 now applies in full to all other Journeys and Work, which in consequence are described as National or Community Regulated Journeys and Work.

The main regulation made under the 1968 Act which relates to PSV drivers' hours is the **Drivers' Hours (Passenger and Goods Vehicles) Modification Order 1971.** This makes a number of useful modifications to the Act for drivers of PSVs.

13.4.1.1 Scope of the Legislation

The drivers of most passenger vehicles, whether or not they are PSVs, are covered by regulations relating to their hours of work and driving.

Different types of vehicles or work are covered by either:-
(a) Community Rules, or
(b) Domestic Rules, or
(c) Neither (a) nor (b) above.

Exemptions The following exemptions apply:-

(i) From EEC Regulations generally:-
(a) Passenger vehicles with 8 or fewer seats (excluding the driver's seat);

---

(71) Public Passenger Vehicles Act, 1981, section 22(5).
(72) The driver must first ask the Traffic Commissioner to reconsider the matter, and may be heard by him. The Magistrate's (in Scotland the Sheriff's) decision is binding on the Traffic Commissioner. Public Passenger Vehicles Act, 1981, section 23.
(73) Section 96 and following sections.

(b) Passenger vehicles on regular services when the route covered by the service does not exceed 50 kilometres (31 miles). A return journey over the same route is not included in determining route length;
(c) Ambulances, specialised medical and rescue vehicles;
(d) Other specialised breakdown vehicles. (The provisions of the Transport Act 1968 apply when used in Great Britain);
(e) Police, Armed Forces, Civil Defence and Fire Brigade vehicles.

(ii) From EEC Rules for journeys within Great Britain:-
(a) Vehicles, which in construction and equipment are suitable for carrying not more than 14 passengers in addition to the driver;
(b) Vehicles undergoing local road test for purposes of repair and maintenance.

(iii) From Part VI of the Transport Act 1968:-
(a) Passenger vehicles, not being PSVs, with 12 or less seats (excluding the driver's seat);
(b) Police, Armed Forces, Civil Defence and Fire Brigade vehicles (74).

It follows from the above that complete exemption from the laws governing drivers' hours applies only to the Armed Forces, Police and Fire Brigades, although non-PSVs with 8 or less seats (excluding the driver's seat) escape coming within the scope of either UK or EEC drivers' hours regulations. Some Ambulances, surprisingly, by inference, come within scope of the Domestic regulations.

13.4.1.2 Journeys and Work

It is necessary to differentiate between vehicles operating under Community rules (i.e. National and International work) and Domestic rules, insofar as permitted hours, breaks and rests are concerned.

Regulations (75) provide that Part VI of the Transport Act 1968 shall not apply to National or International Journeys or Work, so all Community-regulated journeys and work (i.e. categories (ii) and (iii) below) are subject to identical rules. No distinction, therefore, is made between National and International Journeys and Work; both are described as 'Community-regulated'.

The type of vehicle, and the type and location of work performed, determine whether the journey and work are defined as:-
(i) Domestic;
(ii) National;) Both of these are described as 'Community-regulated'.
(iii) International;)

(i) Domestic Journeys or Work These are journeys or work to which the Community rules do not apply, but to which the provisions of the Transport Act 1968, Part VI, do apply. (For example, journeys taken within Great Britain by exempted vehicles, listed under exemption headings (i) and (ii) above). In the main it is regular services under 50 kilometres, and passenger vehicles with 14 or less seats, which continue to be covered by the Transport Act 1968 and the Drivers' Hours (Modification) Order 1971.

(ii) National Journeys or Work These are Community-regulated Journeys or Work in connection with national transport operations in Great Britain.

EEC National vehicles are those used on National work within Great Britain, excluding exempt vehicles covered by the UK Domestic Hours Rules, but including vehicles with more than 14 passenger seats used on:-

---

(74) Drivers' Hours (Passenger and Goods Vehicles) (Modification) Order 1971.
(75) The Drivers' Hours (Harmonisation with Community Rules) Regulations 1978.

(a) Regular services longer than 50 kilometres;
(b) Occasional and shuttle services.

In practice, this means all vehicles with more than 14 passenger seats used on
(a) Private hire;
(b) Excursions and tours;
(c) Long-distance regular express services;
(d) Stage carriage services with such vehicles where the route length exceeds 50 km.

A vehicle does not have to be a PSV to be covered by the requirements.

(iii) International Journeys or Work These are Community-regulated journeys or work in connection with international transport operations to which the Community rules apply.

EEC/AETR International vehicles are vehicles with more than 8 passenger seats used on international journeys. (The requirements under AETR (The European Road Transport Agreement) are substantially the same as those under the EEC rules for International journeys, except that AETR makes no provision for time spent on ferry boats or trains).

Apart from the member states of the Community, the states to which AETR Agreements apply are:-

Austria	Norway	Spain	USSR
Czechoslovakia	Portugal	Sweden	Yugoslavia
East Germany			

13.4.1.3 Meaning of 'Driver', 'Crewman', and 'Regular Service'

Since both the UK and the EEC regulations refer to these, it is important to understand their definitions.

Driver A driver means a person who drives a vehicle, to which the Drivers' Hours provisions of the Transport Act 1968 apply, either in the course of his employment or for the purpose of his own business.

Crewman The EEC Rules define a crew member as the driver, driver's mate or conductor. This definition includes a person carried in the vehicle in order to be available for driving if necessary, and drivers who drive private vehicles, e.g. works coaches.

Regular Service This is a service which provides for the carriage of passengers at specified intervals along specified routes, with passengers being taken up and set down at predetermined stops.

A person driving both goods and passenger vehicles is considered to be a passenger driver when, in a working week, at least half of his time is spent in driving or on work in connection with passenger vehicles, whether the driving and work applies to Community-regulated or Domestic journeys. However, work done with goods vehicles normally counts towards the build-up of a passenger vehicle driver's hours.

13.4.1.4 Meaning of 'Driving', 'Rest', 'Breaks', 'Working Day', 'Working Week', 'Duty' and 'Spreadovers'

Driving Time The Transport Act 1968 (76) defines 'driving' as 'being at the vehicle's controls for the purpose of controlling its movement, whether it is

(76) Section 103(3).

moving or stationary with the engine running.'

The EEC Regulations do not give a precise definition of 'driving'; but by defining a driver as any person who drives a vehicle within the scope of the Regulations (referred to as an 'inscope' vehicle) even for a short period, they in fact include all driving activities. The Transport Act 1968, however, excludes private driving by defining 'driving' as being for the purpose of an owner driver's business, or for the employer's business by an employee driver.

Continuous Driving This is any driving period not interrupted by breaks which satisfy the requirements as to minimum length (see below). Thus, any interruptions for non-driving work or duty must be ignored in calculating the continuous driving units. (Again, explained below).

Breaks and Rest There is a fundamental difference between the 10 hour period of rest, and the half-hour break which a driver is required to have after a continuous period of driving.

This half-hour break is defined as a period during which the driver is able to obtain both rest and refreshment. The break may be on duty, i.e. where the driver is bound by the terms of his employment to remain on or near the vehicle, or the break may be off duty. Rest is always off duty.

Working Day The working day begins at any time of the day or night, and continues until the driver commences his statutory rest period.

It is thus neither a calendar day nor a 24-hour period! The driver is held to be on duty during his working day (subject to the spreadover provisions in the Transport Act 1968 (77)), and off duty during his periods of rest between working days.

Working Week - Domestic Rules The working week is defined as beginning at midnight on Saturday and continuing until midnight the following Saturday (78). It is thus a calendar week. Operators may apply to the Traffic Commissioners to adjust their working week to commence some other day. Thus a municipal operator may request that his working week starts at midnight on Sunday, just before the Monday morning rush.

Working Week - Community Rules The working week is any period of 7 days, during which a crewman must have taken a statutory period of rest. It thus commences after every weekly rest period and is therefore a rolling week of 7 days.

These two different definitions have very important implications, as will be seen when we consider Minimum Weekly rest periods under both Domestic and Community rules.

Duty Time This covers any time spent on duty by an employee in employment which involves driving a passenger vehicle. However, duty is not limited to driving time or to time spent working on, or in connection with, the vehicle or its load. It includes any other time spent acting under the employer's specific instructions.

An employee is not on duty:-
(a) During breaks for rest and refreshment, if during those breaks the driver has no specific duties or responsibilities to discharge for his employer;
(b) If he is carrying out duties for an employer for whom no driving is

(77) Section 96(1).
(78) Transport Act 1968, section 103.

undertaken.

It is important to be able to establish what counts as 'duty' and hence impinges on (cuts into) a driver's rest period.

As far as drivers operating under GB Domestic Rules are concerned, some guidance is available, both from the Department of Transport, and from a number of decided cases. Two important rules should certainly be remembered.

In the first place, it is generally accepted that any activity from which the employer gains a benefit will count as duty.

For example, in Mitchell v. Abbott, 1966, where a driver at the end of his driving period returned to his own home in a relief car, the court held that the time taken to travel home did not count as duty time. He had exercised an option to drive himself home for his own benefit. (He was paid for the time during which he drove home, and his being at home that evening made his scheduling the next day so much easier for his employer!)

Secondly, the fact that payment is made is not conclusive. This is also illustrated by Mitchell v. Abbott, 1966. It is not unusual for employers to pay drivers a premium payment to cover such inconvenience or to pay guaranteed days of certain lengths, or for there to be periods off-duty during spreadovers which are 'paid through'; all of which when totalled amounts to more than the permitted hours.

This, however, does not give evidence of, or constitute, an offence. In Burridge v. Alderton (Traffic Examiner, Eastern Licensing Authority), 1974, the court accepted that the test as to whether a driver is on duty or not, is not payment, but simply whether the driver is under the employer's control.

Further, someone on National journeys and work doing a full-time job for a different employer cannot be said to be taking a daily or weekly rest period, since he cannot freely dispose of his time. He is in attendance at work, even if he does no driving for the other employer (79).

There are, however, many circumstances which are not clear cut, although the following informal guide lines were given to the press by the DTp in the early days of the Transport Act 1968.

On Duty - (a) travelling to a takeover point after 'signing on';
(b) time spent on the 'spare rota';
(c) working for a second employer as a driver;
(d) double-manning;
(e) conducting;
(f) layover time at a terminus;
(g) inspecting or data collecting.

Off Duty - (a) a rostered meal break;
(b) working for a second employer for whom no driving is done;
(c) the mid period of a 'split' or spreadover duty (paid or unpaid);
(d) waiting at their destination for a private hire party, e.g. as a spectator at a football match.

'Spreadovers' The spreadover provisions in the GB Regulations allow a straight-through 16 hours without any compensation, and the limit on duty time under the EEC National and the International EEC/AETR Regulations is (by inference) a

(79) Pearson v. Rutterford and Another, 1982.

maximum spreadover of 14 hours, that is, 24 hours less 10 hours statutory rest.

It is not critical to establish whether breaks themselves are on-duty or off-duty, although it is important to establish when a driver finally goes off-duty for the purpose of computing his period of rest. Thus, if a driver is on an evening stand-by duty for which he is paid, and management decides early in the evening that they are not likely to need that driver again and allows him to sign off without loss of earnings, that driver could be asked to come in the next morning for an early duty commencing in ten hours' time.

Specific Requirements of the Drivers' Hours Regulations

There follows a section-by-section guide to the Hours Regulations. The actual hours and rest regulations are contained in section 96 of the Transport Act 1968, and in EEC Regulation 543/1969.

Maximum Total Daily Driving Time

Under Domestic Rules, no driver may drive a vehicle for an aggregate time in excess of 10 hours per working day (80).

Under Community Rules, the limit on driving is 8 hours, although this period may be extended, not more than twice in any one week, to 9 hours (81).

Continuous Driving Without a Break

Domestic Rules A driver must take a break of at least 30 minutes after he has been driving continuously for 5 hours 30 minutes. Any interruptions from driving (other than breaks as below) for non-driving work can be completely ignored for the purposes of calculating the continuous driving limits.

The Modification Rules 1971 provide an important alternative to the 5 hours 30 minutes duty rule based on continuous driving as follows:-

As an alternative, a driver may work a straight-through 8 hours 30 minutes driving, provided that he takes breaks of non-driving time, e.g. layover at terminals etc., amounting in aggregate to at least 45 minutes, and the last of his driving periods marks either the end of his working day or the start of a 30 minute break.

Thus, this provision (82) means that the 8 hours 30 minutes 'duty' need not be the only work on that day.

Community Rules A driver or crew member must take a break of at least 30 minutes after a continuous driving period of not more than 4 hours. Alternatively he may take in that period two 20 minute breaks or three 15 minute breaks (83).

Maximum Total Weekly/Fortnightly Driving Time

Domestic Regulations There is no Maximum Weekly duty (this term includes driving time), as the 1971 Modification Order provided that this part of the 1968 Transport Act should not apply to drivers of Passenger Vehicles.

Community Regulations Drivers on International and National Journeys and Work

(80) Transport Act 1968, section 96(1).
(81) EEC Regulation 543/1969, Article 7(2).
(82) Drivers' Hours (Passenger and Goods Vehicles) Modification Order 1971.
(83) EEC Regulation 543/1969, Article 8.

must not drive in aggregate for more than 48 hours in a working week or 92 hours per fortnight (84).

Maximum Daily Duty/Spreadover

The maximum daily duty (working day) is precisely prescribed by the GB Domestic Rules (85), but is only inferred by the EEC Community Drivers' Hours Rules insofar as a driver must have a daily rest period of 'ten consecutive hours during the 24 hour period immediately preceding any time of driving or attendance at work.'

Note that a driver or employee on Domestic (86) Journeys and Work is not on duty if he is carrying out duties for an employer for whom no driving is undertaken. On the other hand, however, if he drives an 'inscope' vehicle he is deemed to be on duty within the terms of both the GB and the EEC Regulations.

Where the vehicle is operating under Community Hours Rules and is double-manned by two drivers, spreadovers of 17 to 22 hours are permitted depending on whether or not the vehicle is fitted with a bunk, and providing breaks of at least 6 and 11 hours respectively are taken.

When a vehicle is double-manned by two drivers and operating on International and National Journeys or Work, the following spreadovers are permitted:-

No Bunk 17 hours' spreadover: provided at least 6 hours is completely off-duty and each driver has a daily rest period of not less than 10 consecutive hours preceding any given time when the driver is on duty.

With Bunk 22 hours' spreadover: provided at least 11 hours is completely off-duty. If the facility enables the co-driver to lie down comfortably, the daily rest period may be 8 hours in a 30-hour period preceding any time when the driver is on duty.

The bunk may only be used for a rest period if the vehicle is stationary; otherwise the daily rest must be taken away from the vehicle.

Daily Intervals of Rest

Domestic Rules A driver must have a period of rest (under the Modification Order) of at least 10 hours between any two working days. This can be reduced on 3 days in the working week to 8 hours 30 minutes.

Note, however, that a working day is not a calendar day. It is a period of driving and duty separated by two statutory periods for rest. (It is of course impossible to legislate for a period of rest, since the driver is free to take rest or otherwise).

Community Rules Crew members must have a period for rest of ten hours during the 24-hour period preceding any time driving or on duty. This effectively replaces the GB spreadover provision with one of 14 hours.

---

(84) EEC Regulation 543/1969, Article 7. The week referred to here is a 'rolling week'.

(85) The Drivers' Hours (Passenger and Goods Vehicles) (Modification) Order 1971 provides for a maximum spreadover each working day of 16 hours.

(86) The ruling in Pearson v. Rutterford and Another, 1982, does not extend to drivers on Domestic Journeys and Work unless they undertake mixed (Domestic and National) driving duties.

Thus, whilst the EEC working day is not a calendar day and can be less than 24 hours, it can never, by the definition above, exceed 24 hours. However, under the Community Rules drivers do have the option to select a different pattern of rest periods (but if the '10 hour' system is selected it must not be reduced during the week).

The ten hours may be reduced to nine hours twice a week, provided that on two other periods during the week this is compensated by 11 hours' rest. There is a proviso that if the latter system is selected, the 'Transport Operation' must include a break of not less than 4 hours' uninterrupted duration, or two breaks of 2 hours' uninterrupted duration, during which no work is undertaken. The daily rest period must normally be taken outside the vehicle, but may be taken on a bunk if the vehicle is stationary.

Where there are two drivers, the daily minimum rest period may be reduced to 8 hours in 30 hours where there are bunk beds, or if there is double-manning without bunk beds, to 10 hours in 27 hours. 'Bunk' is not defined except as being 'a bunk enabling crew members not performing any activity to lie down comfortably'.

Minimum Weekly Rest

Domestic Rules The 1968 Act provides that a driver must have a period of at least 24 hours off-duty in a working week (87). However, the Modification Order 1971, which is applicable to GB domestic operations, provides for a rest day of 24 hours every 2 weeks, i.e. each fortnight. The rest period does not have to be a calendar day. It can be taken at the beginning or end of a working fortnight, and can fall partly in one fortnight and partly in the next, provided it is started in the fortnight to which it applies.

The strict interpretation of '24 hours in every two consecutive weeks' means that it is possible to satisfy the requirement by having off day 1 of week 1, with the next rest commencing sometime during day 7 of week 3. Thus, a 'Domestic Driver' may work continuously for a period in excess of 19 days.

Community Rules Under Community (EEC) Drivers' Hours Rules, the driver must have a minimum period of rest each week of 29 hours, and this must be preceded or followed by the minimum daily rest period of 10 hours (or 11 hours or 9 hours, as the case may be). A week is considered as any seven consecutive days.

The week as far as the EEC Regulations are concerned is a rolling week and not, as the UK Regulations state, midnight Saturday to midnight the following Saturday (with provision to vary this for operational reasons with the Traffic Commissioners' consent).

The 29 hours may be reduced to a minimum of 24 hours provided the shortfall is compensated at some other time during the working week (88).

As an alternative, on Community-regulated journeys which are not Regular Services between 1st April and 30th September only, the weekly rest period may be replaced by one of at least 60 consecutive hours in any consecutive 14 days, if preceded or followed by a daily rest period (i.e. 70 hours in total) (89).

---

(87) Transport Act 1968, Section 96(6).
(88) EEC Regulation 543/1969, Article 12.
(89) Ibid., Article 12(3).

Drivers' Hours Regulations can be summarised in tabular form as follows:-

	Domestic Rules	Community Rules
Maximum Daily Driving	10 hours	8 hours
Maximum Continuous Period of Driving Without a Break	*5 hours 30 minutes	4 hours
Minimum Break Period	30 minutes (*)	30 minutes or 2 x 20 minute periods or 3 x 15 minute periods
Minimum Daily Rest	10 hours (reducible to 8 hours 30 minutes on 3 times in any week)	10 hours, reducible to 9 hours twice in 7 days if compensated by two rest periods of 11 hours (+)
Minimum Weekly Rest	24 hours per two weeks	29 hours plus 11 hours
Max. Weekly Driving	-	48 hours
Max. 2-Weekly Driving	-	92 hours

NOTES:-

(*) A through duty of 8 hours 30 minutes may be worked (without break) if the accumulated layover time is at least 45 minutes.

(+) Provided that the preceding duty has included an uninterrupted break of 4 hours, or two uninterrupted breaks each of 2 hours.

13.4.1.6 Miscellaneous Conditions

Ferry Boats or Trains

Under the EEC Community Rules, when a vehicle is transported by ferry boat or train, the daily rest period may be interrupted on one occasion only, so that part of the rest may be taken on board the boat or train and part elsewhere. The period of interruption must be as short as possible and not more than one hour, to allow for embarkation, disembarkation, customs etc.

During both portions of the rest period the crew member must have access to a bunk or couchette, and when so interrupted the daily rest period must be increased by two hours.

Time spent on a ferry boat or train not treated as rest may be regarded as a break.

Emergencies

The Transport Act 1968 defines emergencies very precisely as:-

'Events requiring immediate action to avoid danger to life or health or serious interruption of essential public services (gas, water, electricity and drainage) or of telecommunications or postal services, or in the use of roads or airports, or damage to property'.

There are exemptions from the hours rules to allow drivers to deal with emergencies. The Transport Act 1968 also makes provision for drivers to exceed the permitted hours in the event of what is described as an 'unavoidable delay'. In this case the reasons for the delay should be noted on the driver's tachograph disc or control book, when it is necessary to keep such a document.

However, under the Community EEC Regulations relating to National and International operations, emergencies are much more loosely defined and in a way which more closely resembles the definition of 'unavoidable delay'. An emergency is:-

'An event or danger, or circumstances beyond a crew member's control, under which the crew member will be permitted to disregard the driving time and daily rest requirement to the extent that it is necessary to ensure the safety of the passengers of his vehicle, and to enable him to reach a suitable stopping place or the end of the journey.'

The occurrence and the reason why the requirements were disregarded, must be noted by the crew member in his control book.

Under the AETR Agreement, a driver may extend his continuous and/or daily driving period by up to half an hour in order to reach a suitable stopping place (90).

Part-Time Drivers

Drivers who are covered by the GB Domestic Rules (those mainly on regular services under 50 kilometres, and drivers of small passenger vehicles) who drive on every day of the working week for less than four hours, are not subject to the limitations in the Transport Act 1968 on breaks after continuous driving, or weekly rest period, or maximum daily duty (91).

The Modification Order 1971 allows a driver to exceed 4 hours' driving on up to two occasions per week, and still benefit from exemption as a 'part time driver' on the remaining days on which he drives less than 4 hours on domestic journeys and work.

It should be noted that this '4 hour rule' applies to GB Domestic drivers. It applies over the week as a whole, and is not applicable to the driving of a vehicle on any single day. Nor does the rule apply to the keeping of records. (Records are covered below, section 13.4.2).

On days when the driver does not drive, the rules regarding permitted daily duty and breaks for rest and refreshment will not apply to him during that day (92). It is still necessary, however, for a driver to take an interval of rest of 10 hours (or 11 or 9 as the case may be) between each working day.

Bonus Schemes - Community-Regulated Journeys

Bonus or incentive schemes related to distances travelled are not allowed unless they are of such a nature that they do not endanger road safety.

Mixed UK and EEC Driving

Drivers working partly under EEC Rules (National or International) and partly

---

(90) Article 11 of the AETR Agreement.
(91) Transport Act 1968, section 96(7).
(92) Ibid., section 96(8).

under the Transport Act 1968, must observe three basic principles:-

(a) Time spent driving or on duty under one set of rules may not count as a break period under the other set.
(b) Driving and other duty under the EEC Rules counts towards the limits on driving and other duty under the Transport Act rules.
(c) Driving and other duty under the Transport Act rules counts as attendance at work under the EEC Rules.

We can summarise the combination of rules as follows:-

Driving: Drivers must always be within the 1968 Act's limits (i.e. not more than 5 hours 30 minutes ccntinuous driving, not more than 10 hours daily driving), but at any time when actually driving under the EEC Rules they must also comply with all the applicable requirements on continuous daily, weekly and fortnightly driving, and on daily and weekly rest periods.

Total Daily Duty: Drivers must always be within the 1968 Act's limits on the working day (i.e. 16 hours' duty), but when actually working under the EEC rules they must comply with the EEC requirements on daily and weekly rest periods.

Rest Periods and Breaks: Drivers must at all times be within the 1968 Act rules, but whenever actually operating under the EEC Rules they must observe the more stringent EEC requirements.

Offences, Penalties and Valid Defences

It is an offence to contravene the regulations concerning drivers' hours and rest periods. Contravention renders the driver and his employer liable to a maximum fine of £200 (93).

The Transport Act 1968 introduces two defences in any prosecution for hours offences:-

(a) The driver may show that the contravention was due to unavoidable delay in the completion of a journey. It is essential that delays be entered in the driver's record at the time, to avoid possible misunderstandings.
(b) An employer or transport manager can show that he did not cause or permit the contravention, if he can show that it arose out of the driver working for another employer, and that he - the first employer or transport manager - could not reasonably have been aware of this.

13.4.2 Drivers' Records

The law relating to the requirement for the keeping of Records of Work for Drivers and Crewmen of passenger vehicles can be found in:-

(a) The Transport Act 1968, section 96 and following sections;
(b) The Drivers' Hours (Keeping of Records) Regulations 1976;
(c) The EEC Regulations 543/1969 and 1463/1970;
(d) The Passenger and Goods Vehicles (Recording Equipment) Regulations 1979.

13.4.2.1 Scope of the Legislation

All drivers and crew members of vehicles subject to the Community Hours Regulations are bound by Regulations made under the 1968 Act and the EEC Regulations to keep records of their hours of driving, work and rest. In general, therefore, the drivers and crew members of any passenger vehicle not covered by exemption from the Community Regulations (given generally for all International and National Work and Journeys, or specifically in relation to National Work and

---

(93) Transport Act 1968, section 96(11).

Journeys) are within the scope of the legislation.

In particular, the drivers of passenger vehicles with 8 or more seats (excluding the driver) on International Journeys or Work are required to keep records, and the drivers of passenger vehicles with 14 or more seats on National Journeys or Work are also required to keep records.

The full list of exemptions is given in the introduction to the previous section (13.4.1 Drivers' Hours). Drivers of passenger vehicles on 'Regular Services' under 50 kilometres, are exempt from the EEC Hours and Records Regulation 543/1969 (94).

Note that:-
(a) the vehicle does not have to be a PSV;
(b) although drivers of certain exempt passenger vehicles are not required to keep records, they are still required to observe the hours and rest laws. Where a driver drives both an exempt vehicle and a vehicle covered by the Keeping of Records Regulations, he must keep a record showing the work done on both vehicles during that working day.

Examples of such mixed driving can easily be constructed by matching any vehicle in Column 1 driven part of the day, with a vehicle in Column 2 for the rest of the day.

Column 1: In Scope (Community-Regulated)	Column 2: Out of Scope (Domestic Regulations)
Passenger Vehicle (8 or more seats) on International Work	Passenger Vehicle on Regular Service under 50 kilometres
Passenger Vehicle (14 or more seats) on National Work	8-14 seater 'minibus' on National Work
Note: The record would be a tachograph chart unless the vehicle is used to perform: (i) A Regular Service (of any length - see below); (ii) A Non-Community (AETR) International Journey (see below)	**Note:** If a vehicle having 8-12 seats, and not a PSV, is used on National Work, the 1968 hours regulations do not apply.

13.4.2.2 Forms which the Records can take

The type of record used depends on the type of vehicle being used and the type of work carried out.

There are three possible types of record:-

(i) The Individual Control Book (ICB)

The use of an ICB was (until May 1983) compulsory on all Non-Community International Journeys, i.e. journeys to a country not in the EEC but to which the AETR Agreement applies. However, since that date a tachograph chart may now be used as an alternative.

---

(94) Community Road Transport Rules (Exemptions) Regulations 1978, Section 3.

A complete list of the countries which are signatories to the AETR Agreement is given in the previous section (13.4.1 Drivers' Hours).

When drivers are travelling to these countries, even though they may transit an EEC member state en route, they are subject to AETR rules throughout the journey. These rules (until May 1983) made no provision for the replacement of an ICB by a tachograph disc. However, as with Rosters and Timetables (see below), it is advisable where possible to also use a tachograph on International Journeys, as some member states are reluctant to accept any other records.

Specific Requirements Relating to the Use of the ICB The Community Control Book conforms to the model International Record Book annexed to the AETR Agreement and EEC Regulation 543/1969.

The ICB used for AETR Journeys is the same as the ICB used by goods vehicle drivers under the Drivers' Hours (Keeping of Records) Regulations 1976 (95). The system of daily rest selected should be indicated in box 10a. Drivers to whcm the EEC laws apply who drive on National journeys are exempt by the Drivers' Hours (Keeping of Records) (Amendment No. 2) Regulations 1978 from the need to complete on their ICB:-
Box H - Number of hours driving time;
Box I - Total driving and non-driving time.

(ii) Service Timetable and Duty Roster

EEC Regulation 543/1969 (96) enables operators of Regular Services to draw up a timetable and duty roster as an alternative crew record.

A Regular Service is a passenger or goods service operated at specific intervals along specified routes, and calling at pre-determined stops.

Specific Requirements of the Timetable and Roster Operators must construct service timetables and duty rosters, a copy of that part which relates to the driver's own route being given to him to carry on the vehicle (97). The service timetable must show the pattern of the service in sufficient detail to indicate that it is being followed, and the driver must carry a copy with him whilst at work. The duty roster must show each driver's name, his date of birth, his base and his driving schedule for the current week, the week before and the week after (3 weeks in all). The schedule must show the driver's daily rest periods, his breaks, driving periods and other periods of duty. The duty roster must be signed by the service operator, and the driver must carry his own personal extract whilst at work.

(iii) Tachographs

EEC Regulation 1463/1970 provides for the compulsory fitting and use of tachographs in all passenger vehicles used on Community Regulated Journeys and Work.

The Regulations exempt from the compulsory fitting of tachographs, all Passenger Vehicles used on Regular Services.

---

(95) See Chapter 4 for details of how to complete an ICB.
(96) Article 15.
(97) Since vehicles on Regular Services under 50 kilometres route length are exempt from EEC Regulation 543/1969, their drivers need not keep any records, but they must conform with Domestic Hours Rules.

The EEC Regulations do not concern themselves with whether a vehicle is or is not a PSV, or if the journey is stage, express or contract hire. EEC Passenger Vehicle Work is in fact classified as Regular, Shuttle or Occasional. Only the former does not require tachographs to be fitted and used.

An Occasional Journey (roughly equating to Private Hire, Excursions and Tours) and a Shuttle Service (a particular type of express carriage) do require the use of a vehicle fitted with a tachograph.

Regular Services have already been defined (see above under 'Service Timetable and Duty Roster').

Specific Requirements of the Tachograph Regulations The Regulations define recording equipment as 'equipment installed in a road vehicle to show and record automatically, or semi-automatically, details of the movement of the vehicle and certain working periods of its crew'.

A tachograph is such a recording device, frequently mounted in conjunction with the vehicle's speedometer, and capable of giving the following information:-

(a) Speed trace, showing the speed at which the vehicle is driven at any time during the day;
(b) The distance travelled by the vehicle;
(c) A trace showing when the engine is running;
(d) On the EEC type tachograph, an indicator set by the driver to show time spent by the driver driving, on duty and resting.

The instrument is driven either by the speedometer or electrically from a pick-up in the transmission mechanism. It makes recordings with three styluses on a circular chart which is rotated once every 24 hours by a quartz clock mechanism.

If the vehicle is stationary, the 3 styluses make concentric circular traces. When it moves, two of these automatically record speed and distance: The third stylus is controlled by the 'mode key' which the driver moves to different positions:-

(a) Driving (steering wheel symbol);
(b) Other Period of Work (crossed rectangle symbol);
(c) Rest or Break (bed symbol).

When the vehicle is moving, this third stylus marks a thick band on the chart, instead of a thin line.

Installation and Inspection of Tachographs Tachographs are calibrated on being fitted to the vehicle, and sealed to prevent driver interference. They cannot be used under the regulation unless installed, calibrated and sealed by approved fitters in approved workshops, using a registered mark on the seals (98). An installation plaque must be affixed when the tachograph is fitted showing:-

(a) Mark of approved fitter/workshop;
(b) 'w' - revolutions per kilometre;
(c) 'l' - metres (tyre circumference);
(d) Date of calibration and sealing of tachograph and plaque.

Tachographs must be inspected every 2 years and re-calibrated every 6 years or after any repair likely to affect their reading.

The 2-yearly inspection, due either

(a) 2 years after the date shown on the installation plaque, or
(b) 2 years after the date shown on the 2-yearly inspection plaque,

---

(98) Passenger and Goods Vehicles (Recording Equipment) Regulations, 1979, give the Secretary of State for Transport power to grant such approvals.

must be carried out at a tachograph centre. The instrument must be examined to ensure that it is functioning correctly and accurately (99). The plaque inside the tachograph head will show:-

TWO YEARLY INSPECTION
Centre/Seal No...............
Date........................

The following tolerances, plus or minus the actual reading, are allowed:-

	In Use	On installation/calibration
Distance	4% over 1 km	1% over 1 km
Speed	6 km/hr	3 km/hr
Time	2 minutes/day or 7 minutes/week	

Use of the Tachograph Where a tachograph is fitted, its use in EEC member states is compulsory. Any other form of record is now only permitted where a vehicle is exempt from the requirement to fit a tachograph.

Employer's Responsibilities Employers must ensure that sufficient charts are supplied to employees. Employers should bear in mind that these are personal to the owner, and also allow for the possibility of the charts being damaged or taken by an approved examiner. Owner drivers must also carry an adequate supply.

The employer must collect completed charts not later than 21 days after use, and certify them within another 7 days. Completed charts must be retained for a further 12 months.

Drivers' Responsibilities The driver must insert a disc into the tachograph every day when starting work, and whenever a change of vehicle takes place. The driver must write the following information on the disc:-

(a) his name;
(b) date and place chart begins, and the relevant odometer reading;
(c) date and place chart ends, and the relevant odometer reading;
(d) registration number of vehicle(s) used, and the odometer readings relevant to any vehicle changeover.

Each record chart is used only to record personal work of the driver to whom it is issued, and he must transfer it to other vehicles which he drives during the 24 hour period.

The driver must ensure that the instrument is kept reading (at BST throughout an International Journey) and running (with two charts inserted if double-manning), and transfer his chart to any changeover vehicle, entering in the centre field

---

(99) Passenger and Goods Vehicles (Recording Equipment) (Amendment) Regulation 1984. If the date on the plaque is before 13th June 1983, inspection is not required until the next annual test of the vehicle after the 2nd anniversary of the date on the plaque, or 13th June 1984, whichever is the later.

the registration number and odometer reading of the changeover vehicle.

If a tachograph becomes defective during a journey, the chart must be completed by hand or continued on a temporary chart and attached to the original. The equipment itself must be repaired upon return to base, or en route if return does not occur within 7 days (100).

Used charts must be retained by the driver for at least 2 days (101) when operating within the UK, and for at least 7 days when operating on an International Journey.

The replacement of individual control books by tachograph records is permitted when a driver operates under AETR rules. The use of a tachograph in place of other kinds of records is compulsory in other EEC Member States.

Enforcement Tachographs have to be so designed that it is possible for an authorised examiner to verify, without opening the case, that recordings are being made, and, after opening the case, to read the previous 9 hours' recordings without permanently deforming, damaging or spoiling the sheet.

In the case of damage to a sheet, crew members should attach the damaged sheet to a spare replacement.

When they are away from the vehicle and unable to operate the equipment themselves, the various periods of time must, either manually or by automatic recording or otherwise, be entered on the sheet legibly without damaging it.

Offences, Penalties and Defences Any person contravening the requirements of the regulations is liable on summary conviction (i.e. in the Magistrates' Court) to a fine not exceeding £200.

It is, however, a good defence for an employer if he can prove to the court that:

(a) He has given proper instruction to his employees with respect to the keeping of records, **and**

(b) He has from time to time taken reasonable steps to ensure that these instructions have been carried out.

Reliance on this statutory defence demands that records should be subject to detailed regular checking, and that there should be evidence of action having been taken in relation to faults found (102).

Traffic Commissioners can revoke, suspend, curtail or prematurely terminate an Operator's Licence in respect of records offences, and a driver's Vocational Driving Licence may be revoked or suspended for similar reasons.

It is an indictable offence (a more serious offence triable by judge and jury in the Crown Court) to make or cause to be made a false entry or an alteration to an entry with intent to deceive (103).

---

(100) EEC Regulation 1463/70, Article 18(1)(2).

(101) EEC Regulation 1463/70, Article 17(5)(6) allows the 2-day 'derogation' for member states' drivers on National journeys.

(102) It is a good defence against a charge of not checking records for an employer to show that he complied with the requirements as soon as it was reasonably practical for him to do so (Drivers' Hours (Keeping of Records) Regulations 1976, section 8(4)).

(103) Transport Act 1968, Section 99.

13.4.3 Conclusion

The Hours and Records legislation is complicated, but it is important that drivers and employers co-cperate in seeing that it is adhered to. The consequences of a detected infringement may not stop with a conviction and penalty imposed by magistrates, but may involve the temporary or permanent loss or curtailment of the operator's O-licences. The importance of an owner driver's compliance with the regulations is equally obvious. Finally, it should not be assumed that in cases where records are not required to be kept, it is not necessary to comply with the Domestic drivers' hours and work regulations.

13.5.1.1 Speed Limits

Speed limits may be of two types - general and particular. The former (general) speed limits refer to all vehicles. They can be looked upon as road speed limits, as they apply generally to all vehicles on a specific road, although different roads will, of course, be subject to different speed limits (104). The latter (particular) speed limits apply to particular vehicles depending upon where they are being driven, although where the general and particular speed limits are not the same in any particular set of circumstances, the lower of the two usually applies (except on Motorways) (105).

Example A large coach has a particular speed limit of 50 m.p.h. when driven on a single-carriageway derestricted road, but it can travel on a motorway at 70 m.p.h. However, if it were drawing a trailer it would be limited to 50 m.p.h. on motorways and unrestricted roads.

If a particular vehicle may be defined in more than one way, and hence would appear to be limited by more than one particular speed limit, the lower of the two will apply (106).

General Speed Limits There is now a general overall maximum speed limit of 70 m.p.h. This applies on Motorways, and on 'derestricted' dual carriageways. On single carriageway 'derestricted' roads, traffic is limited to 60 m.p.h. (107).

On restricted roads, a speed limit of 30 m.p.h. exists under the Road Traffic Regulation Act 1967 (108), but higher or lower speeds may be permitted by order. Earlier Acts used the familiar term 'built-up area', but a restricted road has since been redefined (109) as a road provided with a system of street lighting with lamps not more than 200 yards apart, or a road subject to a direction that it shall be a restricted road. In this latter case, in the absence of street lighting, a person cannot be convicted of speeding unless there are the necessary repeater restriction signs, at intervals of not more than 300 yards, or 325 yards where a speed limit of more than 30 m.p.h. is in force.

---

(104) Road Traffic Regulation Act 1967, sections 75 ff.
(105) Road Traffic Regulation Act 1967, schedule 5(13); Motorway Traffic (Speed Limit) Regulation 1974.
(106) Road Traffic Regulation Act 1967, schedule 5.
(107) Unless either (a) lower 'general' speed limits are in force or (b) the particular class of vehicle is already restricted to lower limits. Temporary Speed Limit Order 1977; Temporary Speed Limits (70/60/50 m.p.h.) (Continuation) Order 1978.
(108) Road Traffic Regulation Act 1967, section 71(2).
(109) Road Traffic Regulation Act 1967, section 72. Sub-section (1), as amended by section 61 of the Transport Act 1982, provides for the re-definition of 'restricted roads'; sub-section (2) provides for the imposition of a speed limit on roads equipped with systems of street lighting after 1st July 1957.

Certain roads, while not being restricted within the meaning of the Road Traffic Regulation Act 1967 and the Transport Act 1982 (which redefined the term 'restricted road' as above), may nevertheless be subject to a general speed limit. For example, a Trunk Road is not restricted merely because it has street lighting as specified above, unless this was provided before 1st July 1957 (109). Where such a system of lighting exists, but the road is nevertheless not 'restricted' within the meaning of the Acts, evidence of the absence of any derestriction signs will be sufficient to prove the existence of a speed limit. Sections of restricted road furnished with lighting can be specified as unrestricted, and vice versa (110).

Particular Speed Limits Certain classes of passenger vehicles are subject to particular speed limits, whether used on restricted or 'derestricted' roads. These particular speed limits are determined by the following factors:-
(a) the vehicle's unladen weight;
(b) its seating capacity. In this connection, 'small' means having seats for 8 or less passengers excluding the driver, and 'large' means with seats for more than 8 passengers excluding the driver;
(c) whether drawing trailers, and if so, how many.

Unladen Weight	Seating Capacity	Particular Speed Limit when driven on:-		
		Derestricted Single Carriageway Road	Derestricted Dual Carriageway Road	Motorway (111)
Over 3050kg OR 'large'		50 m.p.h.	60 m.p.h.	70 m.p.h.
3050kg or less AND 'small' (112)		60 m.p.h.	70 m.p.h.	70 m.p.h.
Over 3050kg OR 'large' and over 12 metres long (113)		50 m.p.h.	60 m.p.h.	60 m.p.h.
Passenger vehicles drawing trailers (111)		50 m.p.h.	50 m.p.h.	50 m.p.h.

Source: Motor Vehicles (Variation of Speed Limits) Regulations 1984, schedule 2.

13.6.3 Absolute Liability of PSV Operators

A PSV operator may not limit by contract in any way his liability for death or

(109) Road Traffic Regulation Act 1967, section 72. Sub-section (1), as amended by section 61 of the Transport Act 1982, provides for the re-definition of 'restricted roads'; sub-section (2) provides for the imposition of a speed limit on roads equipped with systems of street lighting after 1st July 1957.
(110) Road Traffic Regulation Act 1967, section 72(3).
(111) Vehicles drawing trailers (including Articulated Buses) are banned from the third lanes on Motorways.
(112) By inference, National speed limits apply.
(113) i.e. Articulated Buses.

personal injury to bus passengers whilst they are being carried in, or boarding or alighting from, a PSV (114). The operator is also under an obligation to report to the Traffic Commissioners any accident or damage to a PSV likely to affect passenger safety (115).

13.7 Weights and Dimensions of PSVs

Maximum Dimensions Length - 12 metres.
("Permit" Minibuses and Community Buses - 7 metres).
Maximum overhang - 60% of wheelbase.
Maximum height - 4.57 metres.
Maximum width - 2.5 metres.

Maximum Axle Weight - 9150 kg.

Maximum Gross Weight (allowing 63.5 kg per passenger):-
2-axle rigid vehicle with wheelbase 3.25 metres - 3.65 metres - 15250 kg)
2-axle rigid vehicle with wheelbase over 3.65 metres - 16260 kg) (116).
3-axle rigid vehicles and trailers (including articulated buses)
with wheelbase at least 3 metres but less than 5.1 metres - 18290 kg)
with wheelbase 5.1 metres or more - 24390 kg) (117).

13.9.2 Certification and Inspection of Fitness of PSVs

PSVs, like Goods Vehicles, are subject to fleet inspections and roadside checks, and similar criteria apply (see sections 9.1 and 9.2). However, PSVs require a Certificate of Initial Fitness (CIF) before ever they are put into operation and, unlike Goods Vehicles, they are not yet subject to Type Approval (118).

Issue of CIFs Certifying Officers are responsible for the issue of CIFs. Applications may be made to the Traffic Commissioners of the Traffic Area in which the vehicle can be inspected (119). Appeals against the Certifying Officer's refusal to issue a CIF can be made within 28 days to the Minister, in writing, stating the grounds of appeal (120).

Inspection of PSVs Certifying Officers and PSV Examiners are empowered to conduct fleet examinations at any reasonable time, carry out roadside checks and if necessary detain vehicles for this purpose (121). This is in addition to the general powers of 'authorised examiners' (122) to 'inspect' or 'test' motor

(114) Public Passenger Vehicles Act 1981, section 29. See also Section 13.11.2.4 in this chapter, also Gore v. Van der Lann, 1967.
(115) Public Passenger Vehicles Act 1981, section 20. See also Section 13.11.2.5 in this chapter.
(116) The maximum weight is reduced to 14230 kg if the main and secondary brakes are less than 50%/25% efficient respectively.
(117) As Construction & Use Regulations 1978, Schedule 7 (See Chapter 7, Fig. 5).
(118) However, a PSV adapted to carry more than 8 passengers may be used either if it has received a CIF, or if a certificate has been issued by the Secretary of State approving the vehicle as a 'Type Vehicle' complying with prescribed conditions of fitness (Public Passenger Vehicles Act 1981, sections 6, 7 and 10).
(119) Using Form PSV417. The fee is £35.
(120) Public Passenger Vehicles Act 1981, section 50(6); PSV (Conditions of Fitness, Equipment, Use and Certification) Regulations 1981, Regulation 56.
(121) Drivers and Conductors must give reasonable assistance to any person authorised to inspect the vehicle, and must not obstruct them in any way. (PSV (Conduct of Drivers, Conductors and Passengers) Regulations 1936).
(122) Including Police Officers authorised in writing by their Chief Constable, and DTp Vehicle Examiners (Road Traffic Act 1972, section 53).

vehicles on the road, subject to the right of the driver to elect for a deferred test (123), or of authorised examiners to inspect a vehicle on premises (124). See Chapter 9, Section 9.2.

Certifying Officers and PSV Examiners may prohibit the driving of an unfit PSV by issuing an immediate or delayed prohibition, and may vary such prohibitions (125). Prohibitions may be removed by the officer/examiner when he is satisfied as to the vehicle's condition. Operators may appeal to the Traffic Commissioners against a PSV Examiner's refusal to remove a prohibition, in order to have the vehicle inspected by a Certifying Officer.

Form PSV414 is used to issue either an immediate or a delayed prohibition. The latter takes effect from a specified date not later than 10 days after its issue. Form PSV414A varies the terms of the prohibition. Form PSV414B permits the movement of a prohibited PSV subject to certain conditions being met, such as:-

(a) Passengers are not carried;
(b) A specified speed is not exceeded;
(c) A trailer is not towed;
(d) A rigid or suspended tow is used;
(e) The vehicle is driven in daylight only;
(f) The vehicle is driven to a specified place.

Form PSV414C is a refusal to remove a prohibition. PSV415 and PSV416 are Removal and Defect Notes respectively.

A PSV414 must be carried on the vehicle at all times until it is removed. The vehicle may only be driven to road-test it (126) or to proceed to a testing station.

Testing of PSVs

All PSVs, irrespective of age, must be tested annually on or before the anniversary of their first registration, either at a DTp HGV Testing Station, or at an operator's premises designated as a PSV Testing Station on the application of the operator.

Items covered by the PSV Test include all those items which are MOT Test requirements applying to all classes of vehicles including PSVs, and additionally those items applicable to PSVs (described as Class VI vehicles in the Motor Vehicles (Tests) Regulations 1981 (127)); these are specified, together with pass/fail criteria, in the PSV Tester's Manual. Checks to ensure that tachographs are fitted and working were added in 1981 (128).

---

(123) The test can be carried out during some 7-day period in the following 30 days, unless a Police Constable considers that the vehicle is unfit to proceed without being tested, or that a test should be carried out immediately because of an accident involving the vehicle.
(124) With the consent of the owner of the vehicle or the premises, or otherwise with 48 hours' notice (72 hours by recorded delivery). No notice is required if the vehicle has been reported in an accident (Motor Vehicles (Construction & Use) Regulations 1978, Regulation 145).
(125) Public Passenger Vehicles Act 1981, sections 7 and 9.
(126) Within 3 miles of the place where it is being repaired.
(127) Also the Motor Vehicles (Test) (Extension) Order 1982, which added this class and provided for annual tests of PSVs to conform with the EEC Roadworthiness Directive 143/1977.
(128) Passenger and Goods Vehicles (Recording Equipment) (Amendment) Regulations 1981.

Testing of other passenger vehicles, and of PSVs not having a CIF

__Class IV__ vehicles other than PSVs, i.e. with seats for 12 or less passengers, are tested annually after their first examination (three years after registration if 8 or less passenger seats, otherwise on the first anniversary of registration) by franchised garages, i.e. MOT Testing Stations. They may also be submitted for testing to County or District Councils authorised by the DTp (Designated Councils).

__Class V__ vehicles, with more than 12 passenger seats (129), and PSVs such as Community Buses and LEA School Buses which have no CIF, must be tested annually on the anniversary of registration, either by a DTp Testing Station, or by a Designated Operator, or by a Designated Council (130).

13.9.3 Construction and Use of PSVs

The Motor Vehicles (Construction and Use) Regulations 1978 apply to PSVs as they do to Goods Vehicles. However, just as there are contained in them regulations which apply only to Goods Vehicles (e.g. relating to the overall length of a Goods Vehicle and its load), so there are other regulations which are specific to PSVs.

__Motor Cars and Heavy Motor Cars__ Passenger vehicles are classed as Heavy Motor Cars when their unladen weight exceeds 3050 kg (3 tons) if they have seats for 7 or fewer passengers, and when their unladen weight exceeds 2540 kg (2.5 tons) in all other cases.

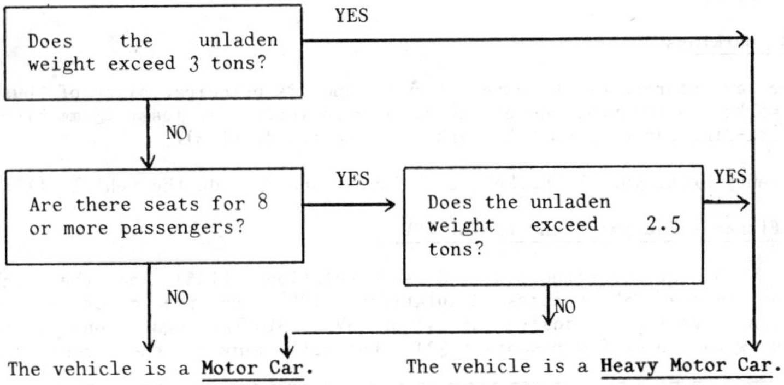

__Tyres__ The tyre regulations apply, but re-cut tyres may be fitted to passenger vehicles which are also Heavy Motor Cars.

__Wearing of Seat Belts__ Drivers and front seat passengers of small passenger vehicles (constructed to carry 12 or fewer passengers) must wear seat belts. Where there is a bench seat at the front for the driver and passengers, the person in the centre need not wear a belt unless the nearside seat with belt fitted is empty, when he must move to that seat and wear the belt. There are

---

(129) This category includes most personnel carriers and minibuses operated under Permits. Large Passenger Carrying Vehicles with 17 or more passenger seats must comply with the PSV (Conditions of Fitness, Equipment, Use and Certification) Regulations 1981 if used under a Section 42 Permit.

(130) A very few franchised MOT Testing Stations are authorised to test Class V vehicles.

certain exceptions for driving instructors and testers, drivers reversing vehicles, and holders of medical exemption certificates. Children under 14 may not travel in the centre front passenger seat unbelted unless every other seat is occupied. Individuals may be fined a maximum of £50 for each offence (131).

Use of Horn (Regulation 118) In addition to the normal rules governing the use of horns (132), horns on PSVs may be used to summon assistance for the driver, conductor or inspector, and to raise the alarm as to theft or intended theft of the vehicle and/or contents.

Trailers (Regulation 135) PSVs may not draw trailers unless they are approved to do so by a Certifying Officer, and such approval will only be given if there is no entrance or exit behind the rear wheels (other than an offside emergency exit). Broken-down vehicles carrying passengers may be towed not faster than 30 m.p.h. with a rigid towbar.

Brakes Requirements as to PSV braking systems, which are additional to the Construction & Use and EEC braking regulations applicable to Motor Vehicles in general, are contained in the PSV (Conditions of Fitness, Equipment, Use and Certification) Regulations 1981. See Section 13.9.5.

13.9.4 Lights

PSVs must carry the same obligatory lights as any other Motor Vehicle, i.e. side and rear lights, stop lights, reflectors, direction indicators, headlamps and high intensity rear fog lamps. PSVs first used before 1st October 1969 may use a single headlamp.

13.9.4.6 Markings

The name and address of the owner of a PSV and his principal place of business must be marked on the nearside of the vehicle in letters at least 25 mm high. The seating/standing capacity must be marked on the inside (133).

All Emergency Exits must be marked, both inside and outside the vehicle (134).

13.9.5 Fitness, Equipment and Use of PSVs

Together with the Construction and Use Regulations (135) and the Lighting Regulations, the PSV Fitness Regulations (136) are the major source of legislation covering the design and use of PSVs. Similar regulations apply to minibuses with 16 or fewer seats (137), but relax many of the 'comfort', as

---

(131) In the case of a child under 14, the driver is held responsible. Transport Act 1982, sections 27/28; Motor Vehicles (Wearing of Seat Belts) Regulations 1982; Motor Vehicles (Wearing of Seat Belts by Children) Regulations 1982.

(132) i.e. to warn of the presence of the vehicle or another moving vehicle. Use of the horn is prohibited between 2330 and 0700, and when the vehicle is stationary.

(133) PSV (Conditions of Equipment, Fitness, Use and Certification) Regulations 1981, Regulation 45.

(134) Ibid., Regulation 24.

(135) Motor Vehicles (Construction and Use) Regulations 1978.

(136) PSV (Conditions of Equipment, Fitness, Use and Certification) Regulations 1981.

(137) Minibus (Conditions of Fitness, Equipment and Use) Regulations 1977 applicable to minibuses used under a Section 42 Permit; Community Bus Regulations 1978 applicable to Community Buses.

opposed to safety, requirements of the PSV Fitness Regulations. The Regulations (136) are complicated in that some of them distinguish between single-deck and double-deck vehicles, and make provision for vehicles first registered some time ago. The following resume assumes a single-deck vehicle being currently registered. Any differences applicable to double-deck vehicles are given in footnotes. No attempt is made to cover additional types such as open-top or half-deck vehicles.

Stability (Regulation 6) PSVs must be able to pass the 'tilt test', i.e. they must not overturn when the surface on which they are standing is tilted through an angle of 35 deg. from the horizontal (138). The vehicle must be loaded with weights of 63.5 kg for every passenger and crew member (139).

Suspension (Regulations 7 and 21) This must be designed to prevent any undue body sway. The body must be securely fixed to the chassis, and trapdoors must be secured against loosening by vibration.

Guard Rails or Body Skirt (Regulation 8) This must extend to within 12 inches (310 mm) of the ground, and to within 6 inches (155 mm) and 9 inches (230 mm) of the rear and front wheels respectively.

Brakes (Regulations 9 and 10) PSVs must meet both Construction & Use and EEC braking standards. In addition no PSV braking system may act through the transmission (140), there must be no universal joint between brake and wheel, and the main brake must be applied with a foot pedal.

Steering (Regulation 12) This must be designed to prevent overlock. Large PSVs adapted to carry over 16 passengers, and if over 7 metres long, must be able to turn within a 12 metre swept circle (141), and outside of a 5.3 metre swept circle (142). The side of the vehicle must not 'cut out' by more than 0.8 metres (143).

Nuts, Bolts and Fastenings (Regulation 11) Nuts, bolts and other fastenings on a PSV must be secured by lock-nuts or split pins, and those securing the steering mechanism must be threaded and have their heads uppermost.

Noise Levels (see also section 9.3.6)

	Maximum dBA on the road
Vehicles manufactured on or after 1st April 1983 and first used on or after 1st October 1983:-	
Under 200 h.p.	85 dBA
200 h.p. and over	88 dBA
Earlier vehicles:-	
With seats for 12 or fewer passengers	87 dBA
With seats for more than 12 passengers	92 dBA

Exhaust Pipes These should outlet at the rear or offside, and close enough to the rear to prevent fumes entering the vehicle. They must be shielded to prevent,

---

(136) PSV (Conditions of Equipment, Fitness, Use and Certification) Regulations 1981, S.I.257.
(138) 28 deg. for a double-deck PSV.
(139) Required on the top deck only of a double-deck PSV.
(140) Excluding 'retarders', which do not 'count' towards the minimum required braking efficiency.
(141) Motor Vehicles (Construction and Use) Regulations 1978, Regulation 9A.
(142) EEC Regulation 36/1980, 'Cut In') Vehicles manufactured after 1st April
(143) 1.2 metres for articulated buses) 1981 and first used after 1st April 1982

as far as possible, any inflammable material dropping on to them.

Electrical Equipment (Regulation 17) Any circuit of more than 100 volts potential must be capable of being isolated by a dual pole switch (144). This must NOT disconnect the PSV's obligatory lights.

Luggage Racks (Regulation 15) These must be designed to prevent any dislodged article falling on the driver. They must be secure and blanked off at the ends.

Internal Illumination (Regulation 16) This must be adequate (145), with a dual circuit so that the failure of either sub-circuit does not black out any deck.

Step Heights (Regulation 20) The maximum height of the lowest entrance step of an unladen PSV is 435 mm (17 inches). The horizontal tread should be at least 225 mm (9 inches) deep (146).

Entrances and Exits (Regulation 24) Every entrance must be on the nearside of a PSV unless fitted with a door controlled by the driver from his seat. The minimum width of entrances is 530 mm (147).

The minimum number of entrances and exits depends primarily on a PSV's seating capacity.

Seating Capacity (Passengers)	Minimum number of Emergency Exits		Position
	Primary Exits (note 148)	Secondary Exits (note 149)	
45 or less	1	1 (note 150)	One on either side
Over 45	1	2 (note 150)	Two of these on opposite sides (note 151)

Single-deck vehicles first used after October 1981 with seats for 17 or more passengers must have at least one emergency exit (150) either
(a) in the rear face of the vehicle; or
(b) in the front face of the vehicle; or
(c) in the roof of the vehicle.

---

(144) Unless one pole is earthed.
(145) Including the top deck of a double-deck PSV.
(146) On double-deck PSVs, stairway 'risers' must be closed in, and there must be a 'landing' 660 mm deep between the top riser and the rear of any seat.
(147) 910 mm (disregarding any stanchion) if it is the only entrance on a double-deck PSV or the entrance most directly serving the stairway.
(148) Leading from the saloon of a single-decker or the lower deck of a double-decker. Minimum dimensions 1.37 metres high and 530 mm wide.
(149) Minimum dimensions 910 mm high and 530 mm wide (1.52 metres x 455 metres if on the top deck of a double-deck PSV).
(150) One of these must also be no smaller than a primary exit.
(151) Any two on the same side must be not less than 3.05 metres apart and within 3.05 metres of (a) the front of the vehicle and (b) the rear of the vehicle.

Emergency exits must **not** be power operated. The means of opening them must be clearly indicated and readily accessible to persons of normal height standing outside the vehicle. (See also section 13.9.4.6, 'Markings'.)

Doors (Regulation 23) These must be fitted with a catch for securing them closed and two devices (handles) for opening them:-
(a) under normal operation by the vehicle owner or his 'agent' (152); and
(b) from the outside.
These must operate with a single movement. Their means and direction of operation must be clearly shown (153).

Doors which might accidentally swing open or closed with the motion of a PSV must also have a device to secure them open.

Power-operated doors (154) must have controls which interlock with the transmission, so that the vehicle cannot move off with the door open.

Access to Exits (Regulation 25) There must be unobstructed access from every seat (155) to at least two exits. No seat may be fitted to any door.

Gangways (Regulation 27 and Schedule 12) Gangways, and passages between gangways and emergency exits, are required on PSVs. Minimum dimensions are prescribed:-
Width: 305 mm (156).
530 mm if within 910 mm of any entrance or exit.
910 mm (157) if leading to lower deck and staircase on a double-deck PSV.
Height: 1.6 metres on single-deck PSVs, and on the lower decks of double-deck PSVs, seating 14 or fewer passengers;
1.77 metres (158) on single-deck PSVs, and on the lower decks of double-deck PSVs, seating more than 14 passengers;
1.42 metres (159) on PSVs with seats for less than 12 passengers.

Seating (Regulation 28) Seats must be securely fixed in position, must be at least 400 mm wide, have a closed back rest (160), and not be any closer than 225 mm to any well or steps. There must be a clear space (leg room) of at least 200 mm in front of every seat, and minimum distances between backrests are prescribed:-
Where both seats face the same way - 610 mm.
Where both seats face each other - 1.6 metres (161).

Folding **'courier seats'** for crew members may be fitted at the front of PSVs (162). Any part protruding into a gangway must automatically retract when the seat is not in use. They must be fitted with armrests.

---

(152) i.e. the driver, conductor or other authorised person.
(153) e.g. 'Pull to Open'.
(154) If they project when opened more than 80mm from the side.
(155) Except (a) seats adjacent to the driver and accessible by a door other than the driver's door, and (b) seats on the top deck of a double-deck PSV accessible to one exit.
(156) Increasing to 355 mm at 765 mm above deck level (i.e. hip height), and 455 mm at 1.22 metres above deck level (i.e. shoulder height).
(157) Or 2 x 455 mm.
(158) 1.72 metres on the top deck of a double-deck PSV.
(159) 1.21 metres within 305 mm of any entrance or exit.
(160) to discourage pickpockets.
(161) 1.37 metres where the PSV has less than 12 seats.
(162) PSV (Conditions of Fitness, Equipment, Use and Certification) (Amendment No. 2) Regulations 1982.

Passenger Protection (Regulation 29) If there is any likelihood of seated passengers being thrown against any entrance or exit, they must be protected by a guard rail or safety glass. Safety glass must be fitted to the windscreen and all windows wholly or partly in front of or to the side of the driver; safety glazing may be fitted to any other window except rear facing windows and windows in doors.

Driver's Accommodation (Regulation 31) The driver must have some means of preventing interior lighting from distracting him, and be able to adjust his seat vertically and horizontally (163). If his seat is accessed from the offside, the entrance must be at least 455 mm wide (164) and, if it is over 690 mm from the ground, a grab rail must be provided.

Ventilation (Regulation 30) This must be adequate, without the need to open windows to demist the windscreen.

Windscreens (Regulation 32) These must be capable of opening to give the driver a clear view ahead, unless there is an adequate demisting and defrosting system.

Bells (Regulation 33) A means of signalling the driver to stop must be provided on PSVs with 12 or more seats.

Equipment (Regulations 35 and 36) Every PSV must carry a fire extinguisher, and every express and contract carriage must carry a first-aid box.

Use of PSVs (Regulations 37-44) No unnecessary obstruction must be caused or permitted to any entrances, exits or gangways; nor must the driver be obstructed unnecessarily. Windows, fittings and seats must be maintained in a clean and serviceable condition. During the hours of darkness, sufficient interior lighting must be provided to illuminate access to all seats and exits, and emergency exit markings. Power-operated doors may only be operated by the driver or by a person authorised by the operator, and then not if the vehicle is in motion. Petrol tanks may not be filled whilst the engine is running, nor may any inflammable or dangerous substances be carried unless in properly designed containers.

Conductors must be carried on PSVs with seats for more than 20 passengers used as stage carriages unless

(a) a Certifying Officer has certified in writing that no conductor is required, or
(b) the PSV is a single-deck vehicle with seats for less than 32 passengers, and the entrance is at the front of the vehicle, and it and the emergency exit are visible to the driver.

13.10.3 PSV Costing

Cost accounting is the recording of actual costs as they occur, and allocation of them to 'cost centres', i.e. resources, as they are employed.

The financial accounts of an operator are necessarily historic, and of little use for monitoring and controlling his operations. They deal only with the costs of the business in aggregate.

---

(163) Where the vehicle has more than 12 seats, the driver's seat must be adjustable over a range of 50 mm in all directions from the position where the backrest is 355 mm and the cushion 200 mm distant from the steering wheel.
(164) With a 455 mm x 530 mm emergency escape if the cab is enclosed.

13.10.3.1 The purpose of a costing system

A costing system is a valuable management tool which can prove immensely useful in both tactical and strategic planning.

In particular:-
(a) it should identify the running and standing costs of vehicles and routes;
(b) it should identify those 'establishment' costs which cannot be directly apportioned to particular vehicles or routes;
(c) it should record revenue by route or 'contract' and compare this with the known cost;
(d) it should be designed to be capable of revealing inefficient operations, e.g. unprofitable routes or contracts, vehicles which are relatively costly to operate, and underutilisation of capital or labour.

13.10.3.2 Methods of Costing

The essential feature of a costing system is that it should record all costs incurred, so that these can be recovered from revenue (165).

A suggested classification of costs is given in the accompanying table (166). Note that depreciation is considered to be a semi-variable cost, i.e. a standing cost which can be attributed to a vehicle but which does not vary in relation to the time or mileage over which the vehicle is operated. Usually the cost of putting a vehicle on the road (167) less its estimated residual or scrap value is spread over the anticipated life of the vehicle, either on the basis of a fixed amount per annum, or by debiting each year a fixed percentage of the written-down value (168). Neither of these methods take account of the effect on vehicle replacement costs of inflation, and more sophisticated methods using current cost accounting principles are often used today. Note that, under section 8 of the Transport Act 1968, a Passenger Transport Authority must include in its revenue estimates all costs properly attributable to revenue, and this includes depreciation of vehicles even where these are leased and not purchased (169).

Allocation of costs Although some costs can be fairly easily directly allocated to individual routes, as they vary according to time or mileage worked on the route, others - including most establishment costs and some semi-variable costs which do not vary in the short term in proportion to time or mileage operated - have to be apportioned on the basis of sharing them amongst the number of vehicles required to provide the current service level at the time of maximum peak demand. Arguably these vehicles still 'stand on the book', even during the 'off peak'.

Allocation of revenue Receipts must also be apportioned route by route, so that costs and receipts can be compared. Comparisons can best be made if all the figures can be 'averaged' on a pence per mile basis, by dividing total route costs and revenue by total route mileage to arrive at a route profit or loss, expressed in pence per vehicle mile.

This is not always entirely straightforward.

Drivers' and conductors' receipts may be obtained from waybills and ticket

---

(165) Also, where appropriate, subsidies.
(166) The system outlined is the CIPFA recommended system.
(167) Less the tyre costs, which are recovered through the running costs.
(168) This 'reducing balance' method is often preferred, since as depreciation lessens as maintenance costs increase, total costs remain fairly constant.
(169) North Western Traffic Commissioners, re SELNEC PTA fares application, 1969.

machine registers, but where more than one route has been operated during a duty, the receipts must be split between these. Revenue from the sale of pre-paid tickets and from subsidies can only be allocated to routes on the basis of a full census or a significant sample of passenger data.

Subsidies may be general (e.g. revenue support or subsidies to enable concessionary fares to be held down), or they may be specific to a particular route which is held to be unremunerative but socially desirable. Obviously the latter type of subsidies are much easier to allocate.

Revenue for private-hire work and regular contracts can usually be set against an ascertainable cost for these, but it must also be borne in mind that, even if such revenue only covers the direct variable costs associated with the job, and makes only a partial contribution to semi-variable costs or establishment costs, it can still be justified on the basis of 'marginal costing'.

Performance Indicators The Code of Practice (170) says that operators should publish nine indicators:-

(1) Operating costs per vehicle mile;
(2) Vehicle miles per employee;
(3) Vehicle miles per vehicle;
(4) Passenger journeys per employee;
(5) Passenger journeys per vehicle;
(6 and 7) Receipts, and subsidies, as a percentage of turnover;
(8) Average fares per mile;
(9) Lost mileage as a percentage of scheduled mileage.

A System of Cost Allocation

Cost Heading		Basis of Allocation
Variable costs		
Drivers' wages, costs and expenses	)	
Conductors' wages, costs and expenses	)	Time
Vehicle servicing	)	
Fuel oil and duty	)	
Tyres	)	Mileage
Hire charges for manned vehicles	)	
Third Party insurance and compensation	)	
Semi-variable costs		
Traffic operational staff costs	)	
Miscellaneous traffic expenses	)	
Maintenance supervisory staff costs	)	Time
Vehicle maintenance (PSVs)	)	
Miscellaneous garage/workshop expenses	)	
Tickets, ticket machines and equipment	)	
Publicity expenses	)	
Vehicle licence duties and fire insurance	)	Peak vehicles
Vehicle depreciation (replacement of PSVs)	)	
Leasing/renting of unmanned vehicles	)	
Fixed costs		
Administration staff costs	)	Time
Education, medical, welfare, benefits	)	

(170) DTp Code of Practice 1981, 'Publication of Information by Operators'; DTp 1981 'Accounting for Bus Operations'.

Rent, rates, fire insurance)
Maintenance, power, light and heating of buildings)
Depreciation of buildings)
Staff cars, vans, lorries) Peak vehicles
Telephone, postage, stationery)
Professional fees, bank charges)
Miscellaneous general expenses)

13.11.2 The PSV Operator and the Law

13.11.2.1 The Carriage of Passengers

An operator who carries passengers for 'hire and reward' incurs a 'common carrier's obligation' to carry for all persons so long as he has room on his vehicle. If he "holds himself out to be, or advertises himself, ready to carry between places on a certain route all persons who accept his published terms and fares, he is not at liberty arbitrarily to reject a passenger who presents himself at the proper place and time and is willing to accept those terms" (171). He must carry anyone who, being in a fit condition, is prepared to pay the fare, and for whom he has room (172). He may refuse, for example, a riotous person likely to cause loss or injury to other passengers, and indeed if he did not do so he could be guilty of negligence in not taking sufficient care of their safety.

However, since a PSV Operator invariably carries under his standard terms and conditions, and thus makes an individual contract with his passenger, he is only under a "private carrier's obligation" extending to liability for his own negligence and that of his servants or agents (compared to the more strict "common carrier's obligation" of being an absolute insurer of their safety). **However**, any attempt to further restrict this liability to a passenger under a contract of carriage is prevented by statute law (see Section 11.2.4). Whereas a fare-paying passenger, in claiming damages for personal injury or death, may rely on his contract of carriage, any passenger may claim also under the general principles of the law of negligence (173).

13.11.2.2 Permitted exclusions

A carrier may contract out of such obligations as those concerning delay and responsibility for passengers' luggage (174).

In the absence of such conditions, he is under a duty to carry the passenger to the destination to which the latter has paid the fare, in a reasonable time, so that delay caused by his negligence or that of his servant renders him liable for damages - but not if the delay could not have been avoided by care and foresight on the operator's part.

An operator is liable for the safety of luggage carried on a PSV, unless the passenger takes charge of it. The operator's duty of care starts when the goods are handed to him, and continues until they are retrieved. He must allow sufficient time in his schedules for passengers to claim their luggage. Goods accompanying passengers which are not for their own personal use, are not "luggage" in the legal sense. In taking charge of "left luggage", carriers are "bailees for reward", and liable for the safety of the luggage. If, for example,

---

(171) Clarke v. West Ham Corporation Transport Department, 1909.
(172) Wilkie v. London Passenger Transport Board, 1947.
(173) Kelly v. Metropolitan Railway, 1895.
(174) The Secretary of State may make regulations relating to passengers' luggage (Public Passenger Vehicles Act 1981, section 60(j)).

they allow another person access to the luggage without establishing that person's right to do so, the operator is liable for any loss occurring (175).

The extent to which carriers may exclude their liability is subject to provisions in recent consumer legislation (176).

13.11.2.3 The Contract of Carriage

The contract of carriage is made between the operator and the passenger when the latter, having invited the operator to stop and pick him up, by his conduct in entering the vehicle, indicates an assumed willingness to accept the offer which the operator makes to carry him, and in consideration to pay the fare (172). The contract becomes manifest when the passenger buys a ticket, and the parties' mutual rights and obligations are governed by the conditions under which the ticket is issued. The courts have over the years decided a number of "ticket cases", the import of which is as follows:-

(1) If a ticket refers to published conditions, the passenger is held to be bound by these (177) so far as they satisfy "the requirements of reasonableness", and do not unreasonably take away the affected consumer's rights as to the performance of the contract (176).

(2) A passenger who knowingly accepts a cheap ticket (below the fare scale which statutory undertakings must display on the vehicle) receives "sufficient notice" of the possibility of there being conditions attached (178).

(3) In general, if a defence based on a notice contained in a ticket is to succeed, there should be a genuine attempt to draw this to the attention of the recipient and make him aware of the precise nature of the limitations (179). The defendant should be able to show that the plaintiff could reasonably be expected to have known of the limitations, and by his conduct accepted them (180).

13.11.2.4 Limitation of Liability

An operator must, so far as human care and foresight can do, provide for the safety of his passengers, whether they are carried under a contract of carriage or gratuitously. The extent to which he can limit or exclude this liability is limited by statute (181). An operator may not limit **by contract** his liability for death or personal injury to his passengers whilst they are being carried in, or boarding, or alighting from, a PSV. Changes in the law of third party insurance (182) which extend cover to all passengers in Motor Vehicles, to some extent overtake the previous caveat.

It is still possible for an operator to exclude his liability to a gratuitous or non fare paying passenger (e.g. a child under age, or a passenger on a "free

---

(172) Wilkie v. London Passenger Transport Board, 1947.
(175) Alexandre v. Railway Executive, 1951.
(176) Unfair Contracts Terms Act, 1977.
(177) Thompson v. London, Midland and Scottish Railway, 1930.
(178) Penton v. Southern Railway, 1930, and also note (169) on previous page. There is "the doctrine of the Fair Alternative" - the passenger may still elect to travel at full fare and standard conditions. Also additional conditions may attach to a concessionary fare, e.g. that it is not available at peak hours.
(179) Thornton v. Shoe Lane Parking, 1971.
(180) Bennet v. Tugwell, 1971. The case involved a "cab notice".
(181) Public Passenger Vehicles Act 1981, section 29.
(182) Road Traffic Act 1972, section 143.

pass"), provided that the passenger's permission to travel can be construed as a licence and not as a contract (183). However, if the circumstances under which a pass is applied for and obtained contain the elements of a contract, any exclusions therein are void by statute (181) (184).

13.11.2.5 Negligence

Negligence amounts to a failure to exercise the ordinary duty of care which a carrier owes to his passenger, or to any other person whom he could reasonably have foreseen could suffer from his actions or those of his servant. Where an accident occurs as a result of the breach of his statutory duty of care in not maintaining his vehicle, he is clearly negligent. He has an objective standard of care as regards his vehicles, amounting to a warranty that they are as safe to travel in as anyone's reasonable care can make them (185).

Irrespective of any considerations under contract law, gratuitous passengers have the same rights to safety as passengers who have paid the fare. The operator must also exercise reasonable care for the safety of all visitors to the premises which he occupies - this includes those persons either invited or licensed to call there (186).

On the other hand, the only duty which the operator owes to a trespasser is that he must not act in reckless disregard of the trespasser's safety. The occupier must exercise a higher standard of care in those places where active operations are carried out (e.g. bus stations and depots) than where this is not so (e.g. at office locations). In particular, a very onerous duty of care exists anywhere where young children might be expected to play (187).

Where the plaintiff in a negligence case has, by his own acts or omissions, contributed to the loss which he has sustained, damages may be apportioned between the parties in proportion to their respective degrees of negligence (188).

The burden of proof in negligence cases is usually on the plaintiff, but where, as for example in the case of a bus crash, it would be unreasonable to expect the plaintiff to prove the defendant's seemingly obvious negligence, the burden may be reversed, unless the defendant can refute the allegation of negligence (189).

Where an employee is carrying out those duties for which he was employed under a contract of employment, then the employer himself becomes liable for the consequences of the employee's actions (190). In the same way, an employee injured whilst performing his duties would obtain damages from his employer (191). But where an employee is performing some action for which he was not employed, the employer is not liable for the consequences (192).

---

(181) Public Passenger Vehicles Act 1981, section 29.
(183) Clarke v. West Ham Corporation Transport Department, 1909.
(184) Gore v. Van der Lann, 1967 (ex parte Liverpool Corporation Transport Dept).
(185) Barkway v. South Wales Transport, 1950.
(186) Occupiers' Liability Act 1957; Health and Safety at Work etc. Act 1974.
(187) Herrington v. British Railways Board, 1971.
(188) Law Reform (Contributory Negligence) Act 1945; Oliver v. Birmingham & Midland Motor Omnibus Co., 1933.
(189) The doctrine of "res ipsa loquitur" (the thing speaks for itself).
(190) Limpus v. London General Omnibus Co, 1862. The driver raced his bus on the highway, in disregard of his employer's express instructions.
(191) Videan v. British Transport Commission, 1963.
(192) Beard v. London General Omnibus Co, 1900, in which a conductor negligently drove the bus (distinguish from Limpus v. London General Omnibus Co, above).

In a recent case, a driver could only have prevented an accident by leaving the driving seat and manually opening the doors. A passenger's clothes were trapped in the doors and caused her to be dragged along and injured. A charge of failing to take reasonable precautions was dismissed (193).

13.11.6 Luggage and Lost Property

The Minister may make regulations regarding the carrying of luggage on PSVs (194) and lost property (195).

Lost Property Regulations (196)

Anyone who finds lost property on a PSV must hand it to the driver or conductor, who in turn must ensure that the property is handed in within 24 hours to the operator. Alternatively, if such lost property is claimed on the vehicle and the driver/conductor is satisfied that the claimant is the owner, it must be returned to him forthwith. The facts must then be reported by the driver/conductor to the operator, including the name and address of the owner and a description of the property. No charge is to be made in these circumstances to the owner.

The driver/conductor is required to search the vehicle for lost property before and on termination of any journey.

The operator is required to keep records of lost property, and to keep it safe until it is claimed by the owner or disposed of. Official documents must be returned to the issuing body. The owner must be notified if his name and address is ascertainable. Records must be kept for a minimum period of 12 months.

The operator must return lost property to the owner, so long as he is satisfied in this respect, on payment of a prescribed fee. The owner must give his name and address.

The scale of fees is set out in Schedule 2 of the Regulations (196).

If property is unclaimed within one month, the operator is permitted to dispose of it by sale, although the operator may extend the claim period to three months as he thinks fit.

Any documents (including official documents) which it has not been possible to return, are to be disposed of by the operator after three months.

Proceeds of the sale of lost property, and money which has been lost, may be disposed of as the operator sees fit.

Perishable goods may be disposed of within 48 hours if not claimed, or if such property becomes objectionable, it may be disposed of at any time.

Refusal by the owner to pay required fees enables the operator to treat the lost property as though unclaimed.

---

(193) Edwards v. Rigby, 1980, Divisional Court. The charge was brought under section 4 of the PSV (Conduct of Drivers, Conductors and Passengers) Regulations 1936.
(194) Public Passenger Vehicles Act 1981, section 60j.
(195) Ibid., section 60k.
(196) The PSV (Lost Property) Regulations, 1978; similar regulations apply to the London Transport Executive.

13.11.7 Carrying Capacity of PSVs

Regulations (197) may be made (198) relating to the carrying capacity of PSVs, the number of standing passengers and the markings on vehicles showing these capacities.

On Stage Carriages, not more than 8 standing passengers, or one-third of the seating capacity, may be carried on a single-deck PSV or on the lower deck of a double-deck PSV. This is permitted only during 'peak hours', or when hardship would otherwise be caused. No standing passengers may be carried on the upper deck of a double-deck PSV.

Where only children, not over 15 years of age, are carried, three seated children may be reckoned as two passengers.

Traffic Commissioners may prohibit or restrict the carriage of standing passengers on any service. Standing passengers are generally permitted only on Stage Carriages, although the Traffic Commissioners may authorise them to be carried on Express Carriages. Standing passengers are not permitted on Contract Carriages.

The Certifying Officer may certify that a vehicle may carry a specified number of standing passengers when it is specially constructed or adapted (e.g. a 'standee bus'); permission may be granted in respect of specified routes or areas.

13.11.8 Conduct of Drivers, Conductors, Inspectors and Passengers

Regulations (199) may be made (200) regarding the conduct of drivers, conductors, inspectors and passengers. They may provide for such matters as

(a) payment of fares;
(b) production, acceptance or surrender of tickets;
(c) production of a licence (for the purpose of endorsement);
(d) removal from a PSV by the crew (or at their request by a police constable) of any person infringing the regulations, and the requirement for any passenger whom the crew reasonably suspect of infringing the regulations to give his name and address.

A police constable may arrest without warrant a person whom he has reasonable cause to suspect has not complied with the regulations, and either refuses to give his name and address, or does not satisfy the constable that the name and address is correct (201).

Conduct Regulations

Drivers and conductors must:-

(a) not smoke on a vehicle during a journey or with passengers on board;
(b) take reasonable precautions for the safety of their passengers;
(c) inform passengers of destination, route or fare, and not wilfully deceive in this matter;
(d) give reasonable assistance to any person having authority to inspect the

---

(197) The PSV and Trolley Vehicles (Carrying Capacity) Regulations 1954, as amended (S.I.s 472/1958, 674/1966 and 76/1980) remain in force.
(198) Public Passenger Vehicles Act 1981, section 26.
(199) The PSV (Conduct of Drivers, Conductors and Passengers) Regulations 1936, amended 1975.
(200) Public Passenger Vehicles Act 1981, sections 24 and 25.
(201) The PSV (Arrest of Offenders) Act 1975 is consolidated in this part of the Public Passenger Vehicles Act 1981.

vehicle, and not obstruct them in any way;

(e) give particulars of their licence, and employer's name and address, to any police constable or person having reasonable grounds for requiring it;

(f) ensure that indicators, where provided, are correctly displayed;

(g) not use unnecessary force if called upon to remove a passenger as above.

A conductor must not distract the attention of the driver whilst he is driving, without good cause.

Passengers must not:-

(a) use obscene or offensive language or behave in a disorderly manner;

(b) enter or remain on the vehicle when asked by the crew to leave on the grounds that the vehicle is full, or is not allowed to pick up or set down at the point in question;

(c) travel on the upper deck without occupying a seat;

(d) cause injury or discomfort to any passenger;

(e) interfere with the working of the vehicle in any way, or spit in it or from it, or soil or defile it in any way, or throw anything out of it;

(f) distract the driver's attention whilst the vehicle is in motion, or speak to him other than to give directions as to stopping; or give any signal which the driver might interpret as a signal from the conductor to start;

(g) distribute printed matter for the purpose of advertising, or, except in the case of a Contract Carriage, offer any article for sale;

(h) smoke when prohibited;

(i) damage or alter fare tables, indicators or advertisements;

(j) cause any annoyance by singing or shouting, or operating a noisy instrument.

A passenger on a Stage or Express Carriage must not, without reasonable excuse, leave or attempt to leave the vehicle without having paid the fare for the journey.

A passenger must, if asked, declare his journey. He must, unless he has a valid ticket, pay his fare, and leave the vehicle at the end of the journey for which he has paid (or pay for continuing the journey), and if required surrender his ticket. He must produce his ticket or pay his fare if required to do so by an authorised person.

Appendix 1 ~ Table of Statutes

(In this Table, the Statutes, Statutory Instruments, EEC Regulations etc are set out in one list in date order. References to each Statute as a whole are shown first, followed by any detailed references to sections, schedules, articles etc within each Statute. The following abbreviations are used:- s - section, ss - sections, sch - schedule, reg - regulation, ff - following section(s). Where known, the S.I. number of each Statutory Instrument is shown in brackets after its title, e.g. S.I.1612/1954).

Public Service Vehicles (Conditions of Fitness, Equipment, Use and Certification) Regulations 1981 (S.I.257/1981) 173, 197-202
reg6 - 199; reg7 - 199; reg8 - 199; reg9 - 199; reg10 - 199; reg11 - 199; reg12 - 199; reg15 - 200; reg16 - 200; reg17 - 200; reg20 - 200; reg21 - 199; reg23 - 201; reg24 - 198, 200; reg25 - 201; reg27 - 201; reg28 - 201; reg29 - 202; reg30 - 202; reg31 - 202; reg32 - 202; reg33 - 202; regs35/36 - 202; regs37-44 - 202; reg45 - 198; reg56 - 195; sch12 - 201
Transport Act 1981 .... 30, 31, 97
Employment Act 1982 .... 138, 153, 155-157
s17 - 155; s58 - 154; s59a - 154
Finance Act 1982, s8 28
Goods Vehicles (Plating and Testing) Regulations 1982 (S.I.1478/1982) 109, 113
sch2 - 109
Industrial Training Act 1982 158
Motor Vehicles (Construction and Use) (Amendment) (No. 7) Regulations 1982 (S.I.1576/1982) 16, 73, 92
Motor Vehicles (Test) (Extension) Order 1982 (S.I.1550/1982) 196
Motor Vehicles (Type Approval for Goods Vehicles) (Great Britain) Regulations 1982 (S.I.1271/1982) 107
Motor Vehicles (Wearing of Seat Belts) Regulations 1982 (S.I.1203/1982) 198
Motor Vehicles (Wearing of Seat Belts by Children) Regulations 1982 (S.I.1342/1982) 198
Public Service Vehicles (Conditions of Fitness, Equipment, Use and Certification) (Amendment No. 2) Regulations 1982 (S.I. 1058/1982)..... 201
Social Security and Housing Benefits Act 1982, s39 .... 162, 163
Transport Act 1982 194
ss27/28 - 198; s52 - 13-17, 20; s61 - 56, 193; sch4 - 13, 14, 16, 17, 20
Motor Vehicles (Construction and Use) (Amendment) (No. 2) Regulations 1983 (S.I.471/1983) 93
Motor Vehicles (Construction and Use) (Amendment) (No. 3) Regulations 1983 (S.I.932/1983) 92
Road Vehicles (Marking of Special Weights) Regulation 1983 (S.I. 910/1983) 85
Social Security Adjudications Act 1983 162
Goods Vehicles (O-Licences, Qualifications and Fees) Regulations 1984 17, 20, 87
s3 - 12; s8 - 22; s9 - 14; s10 - 22, 24; s11 - 13; s17 - 13; s32 - 22; sch5 - 12; sch6 - 15; sch6(5) - 22
Metropolitan Traffic Area (Transfer of Functions) Order 1984 (S.I. 31/1984) 175
Motor Vehicles (Variation of Speed Limits) Regulations 1984 (S.I. 325/1984) 56
sch2 - 194
Passenger and Goods Vehicles (Recording Equipment) (Amendment) Regulation 1984 (S.I.144/1984) 191
Road Vehicles Lighting Regulations 1984 (S.I.81/1984) 97
s4(4) - 100; s6(6) - 106; s16 - 98, 100; s19 - 103; s20 - 98; s20(1) - 102; s20(4)(c) - 100; s22 - 98; s22(2)(b) - 100; sch2 - 98, 106; sch2(1) - 98; sch2(8) - 102; sch7 - 102; sch7(2) - 102; sch8 - 102; sch9 - 103; sch10 - 98; sch11 - 100; sch16(1) - 103; sch18 - 106

Appendix 2 ~ Table of Cases

(In this Table, the cases appear in chronological order)

Coggs v. Bernhard, 1704 133
Stephenson v. Hart, 1828 133
Johnson v. Midland Railway, 1849 131
Hadley v. Baxendale, 1854 133
Somes v. British & Empire Shipping Co., 1860 133
Limpus v. London General Omnibus Co., 1862 ..... 135, 207
Nugent v. Smith, 1872 131
Hoare v. Great Western Railway, 1877 133
Stephens v. Great Western Railway, 1885 133
Kelly v. Metropolitan Railway, 1895 205
Beard v. London General Omnibus Co, 1900 207
Clarke v. West Ham Corporation Transport Department, 1909 ..... 205, 207
Barnfield v. Goole & Sheffield Transport, 1910 133
Gill v. Carson & Nield, 1917 59
Penton v. Southern Railway, 1930 206
Thompson v. London Midland & Scottish Railway, 1930 206
Bryant v. Marx, 1932 59
Dawson v. Winter, 1932 63
Donoghue v. Stephenson, 1932 134
Oliver v. Birmingham & Midland Motor Omnibus Co, 1933 207
Monk v. Warbey, 1935 65
Trickett v. Queensland Insurance, 1936 65
Peters v. General Accident Co, 1938 64
Paterson v. Burnet, 1939 27
McLeod v. Buchanan, 1940 60
NFU Mutual Insurance v. Dawson, 1941 65
Gifford v. Whittacker, 1942 97
Bourhill v. Young, 1943 62
Harvey v. Road Haulage Executive, 1945 134
Kerr v. McNeill, 1945 31
Wilkie v. London Passenger Transport Board, 1947 ..... 132, 205, 206
Harding v. Price, 1948 62
Barkway v. South Wales Transport, 1950 ..... 134, 207
Alexandre v. Railway Executive, 1951 ..... 133, 206
Andrews v. Kershaw, 1952 ..... 84, 97
Carpenter v. Campbell, 1953 ..... 27, 60
James v. Smee, 1954 60
Green v. Burnett, 1955 97
Sprake v. Tester, 1955 58
Soloman v. Durbridge, 1956 59
Hart v. R., 1957 31
Elliot v. Gray, 1959 60
National Coal Board v. Gamble, 1959 61
Clifford Turner v. Waterman, 1961 58
Newberry v. Simmonds, 1961 26
Flower Freight Co v. Hammond, 1962 27
McLeod v. Wajkowska, 1963 58

Videan v. British Transport Commission, 1963 207
Clarke v. National Insurance Association, 1964 65
Holder v. Walker, 1964 58
Liverpool Corporation v. Roberts and Marsh, 1964 65
Mitchell v. Abbott, 1966 .... 38, 181
R. v. Industrial Injury Commissioners (ex parte AEU), 1966 162
Gore v. Van der Lann, 1967 (ex parte Liverpool Corporation Transport Department) 195, 207
Crutchley v. British Road Services, 1968 132
Potter v. Gorbould, 1968 38
North Western Traffic Commissioners re SELNEC PTA fares application, 1969 203
Scotflow v. Eastern Licensing Authority, Transport Tribunal, 1969 21
Transport Tribunal re B H King & Son (Transport) Ltd, 1970 15
Ready Mixed Concrete (East Midlands) Ltd v. Yorkshire Traffic Area Licensing Authority, Transport Tribunal 1970. ..... 12, 60, 88, 135
Bennett v. Tugwell, 1971 .... 135, 206
Herrington v. British Railways Board, 1971 .... 134, 207
Kelman of Turriff, appeal before Transport Tribunal, 1971 14
Bullen, appeal before Transport Tribunal, 1971 15
Stevens v. R., 1971 79
Thornton v. Shoe Lane Parking, 1971 .... 135, 206
Gillespie Bros & Co Ltd v. Roy Bowles Transport Ltd., 1973 132
Burridge v. Alderton (Traffic Examiner, Eastern Licensing Authority), 1974 .... 38, 181
A Cash & A McCall v. North Western Traffic Area Licensing Authority, Transport Tribunal, 1976 18
North Western Licensing Authority re G J Pilling, 1978 21
West Midlands Licensing Authority re G Seedhouse Transport Ltd, 1978 21
North Western Licensing Authority v. Tootle's Textiles, 1978 21
North Western Licensing Authority re Go Fast Shipping, 1979 21
Scot Bowyers Ltd v. South Wales Licensing Authority, Transport Tribunal 1979 21
Boldizer v. Knight, 1980 63
Britton v. Loveday, 1980 63
Concorde Transport v. Metropolitan Licensing Authority, 1980 46
Edwards v. Rigby, 1980 208
Lakenby v. Brown of Wem Ltd, 1980 46
McCathy v. Smith, 1980 159
McQuaid v. Anderson, 1980 30
McShane v. Express Newspapers, House of Lords, 1980 139
Payne v. Holland, 1980 98
Rumbles v. Poole, 1980 30
Smith v. McCathy's (Wembley) Ltd., European Court of Justice, 1980 159
Alcock v. G C Criston Ltd, 1981 47
Brindley v. Willett, 1981 79
Pickfords Removals v. Metropolitan Deputy Licensing Authority, 1981 21 37, 47
Pearson v. Rutterford and Another, 1982 .... 39, 181, 183
Gibson v. Nutter, 1983 29
Hawkins v. Harold A Russett Ltd, 1983 79
Kammac Trucking and North Western LA, Transport Tribunal 1984 15

Appendix 3 ~ Syllabuses

SYLLABUS FOR THE CERTIFICATE OF PROFESSIONAL COMPETENCE REQUIRED BY THE GOODS VEHICLE OPERATORS (QUALIFICATIONS) REGULATIONS 1977 (as amended)

AIMS:
1. To provide the knowledge necessary to the road haulage operator in the safe and satisfactory performance of his duties in accordance with the EEC Directive and consequential GB legislation.
2. To ensure that the road haulage operator is aware of the relevant sources of information to assist him in up-dating his knowledge as necessary.

LENGTH OF COURSE:
The content of this syllabus is based on a course of 65 hours' direct teaching or its equivalent.

TOPICS At the end of the course the candidate should be able to:

A. ROAD SAFETY (20 hours)

A.1 Driver's Hours and Records (7 hours)
(Questions on exemptions will not be asked)

1.1 Basic hours' requirements (3 hours).

1.1.1 Describe the scope of the regulations regarding Domestic/ National/International/Mixed journeys and the types of vehicle to which the regulations apply.

SYLLABUS FOR THE CERTIFICATE OF PROFESSIONAL COMPETENCE REQUIRED BY THE PUBLIC PASSENGER VEHICLES ACT 1981

AIMS:
1. To provide the knowledge necessary to the road passenger transport operator in the safe and satisfactory performance of his duties in accordance with the EEC Directive and consequential GB legislation.
2. To ensure that the road passenger transport operator is aware of the relevant sources of information to assist him in up-dating his knowledge as necessary.

LENGTH OF COURSE:
The content of this syllabus is based on a course of 60 hours' direct teaching or its equivalent.

TOPICS At the end of the course the candidate should be able to:

A. ROAD SAFETY (17 1/2 hours)

A.1 Driver's Hours and Records (7 hours)

1.1 Basic hours' requirements (3 hours).

1.1.1 Describe the scope of the regulations regarding Domestic/ National/International/Mixed journeys and the types of vehicle and operation to which the regulations apply.

Goods Syllabus

1.1.2 State and apply, where applicable, the requirements of these regulations for:
(a) Driving time.
(b) Duty time.
(c) Breaks and rest periods.
(d) Working day/week.
(e) Exemptions and modifications.
(f) Unavoidable delays - emergencies.
(g) Offences and penalties.
(h) Employers' obligations.

1.2 Hours of Work Records (4 hours) (Record Sheets).

1.2.1 State and apply the basic requirements of the regulations relating to:
(a) Vehicles, journeys and work covered.
(b) Type of record and method of completion: Individual control book.
(c) Owner driver.
(d) Signature.
(e) Examination.
(f) Unavoidable delays.
(g) Employers' obligations - checking and custody.

1.3 Hours of Work Records (Tachographs).

1.3.1 State the requirements for the use of tachographs as specified by current regulations.

1.3.2 State the threshold vehicle weight at which required.

Passenger Syllabus

1.1.2 State and apply, where applicable, the requirements of these regulations for:
(a) Driving time.
(b) Duty time.
(c) Breaks and rest periods.
(d) Working day/week.
(e) Exemptions and modifications.
(f) Unavoidable delays - emergencies.
(g) Offences and penalties.
(h) Employers' obligations.

1.2 Hours of Work Records (4 hours) (Record Sheets).

1.2.1 State and apply the basic requirements of the regulations relating to:
(a) Vehicles and types of operation covered.
(b) Type of records and method of completion for regular and other work:
(i) Timetable.
(ii) Service roster.
(c) Timetables/service rosters which must be carried.
(d) Signature.
(e) Examination.
(f) Unavoidable delays.
(g) Employers' obligations - checking and custody.
(h) Duty schedules/rosters, work allocation records.

1.3 Hours of Work Records (Tachographs).

1.3.1 State the requirements for the use of tachographs as specified by current regulations.

1.3.2 State the circumstances in which required.

Goods Syllabus

1.3.3 State the procedure for having an approved tachograph fitted and the action to be taken in the event of failure.

1.3.4 Read and interpret the recorded information.

1.3.5 Identify recordings resulting from faults and misuse.

1.3.6 Explain the requirements for calibration and sealing.

1.3.7 Describe the requirement for retaining records.

A.2 HGV Driving Licences (2 hours)

2.1 State general requirements for HGV driving licences.

2.2 State groups and classes of HGV licence.

2.3 State age limits for HGV licence.

2.4 State general requirements of Young HGV Drivers' Training Scheme.

2.5 State occasions necessitating production of an HGV licence.

2.6 State reasons for suspension and/or revocation of an HGV licence.

2.7 Define employers' responsibility and liability.

Passenger Syllabus

1.3.3 State the procedure for having an approved tachograph fitted and the action to be taken in the event of failure.

1.3.4 Read and interpret the recorded information.

1.3.5 Identify recordings resulting from faults and misuse.

1.3.6 Explain the requirements for calibration and sealing.

1.3.7 Describe the requirement for retaining records.

A.2 PSV Driving Licences (2 hours)

2.1 State general requirements for PSV driving licences.

2.2 Describe the object and scope of the PSV driving test and dual test.

2.3 State categories and classes of PSV licence (including circumstances in which an HGV may be driven with a PSV licence), types of PSV to which they apply and the issuing authorities.

2.4 State age limits.

2.5 Explain general conditions about signing licences and wearing of badges.

2.6 State occasions necessitating production of licence.

Goods Syllabus

A.3 General Traffic Regulations (5 hours)

3.1 Speed limits (1/2 hour).

3.1.1 State the general speed limits applying to the various categories of roads in Great Britain.

3.1.2 State the particular speed limits imposed on various types of goods vehicles.

3.1.3 State the codes of conduct regarding driving in fog, limited visibility or other adverse conditions.

3.2 Parking, waiting, loading and unloading (2 hours).

3.2.1 State the general requirements and regulations governing:
(a) Clearways.
(b) Parking on and overtaking at pedestrian crossings.
(c) Obstruction and dangerous positions.
(d) Stopping to load and unload.
(e) Stopping at banned kerbs/protected bus stops.
(f) Parking at night on roads.
(g) Overnight parking restrictions.
(h) Bus priority schemes.

3.2.2 Outline powers of bodies responsible for traffic control and enforcement.

Passenger Syllabus

2.7 State reasons for suspension and/or revocation.

2.8 State procedure for renewal of licences.

A.3 General Traffic Regulations (5 hours)

3.1 Speed limits (1/2 hour).

3.1.1 State the general speed limits applying to the various categories of roads in Great Britain.

3.1.2 State the particular speed limits imposed on various types of passenger vehicles.

3.1.3 State the codes of conduct regarding driving in fog, limited visibility or other adverse conditions.

3.2 Parking, waiting, loading and unloading (2 hours).

3.2.1 State the general requirements and practice governing:
(a) Clearways.
(b) Parking on and overtaking at pedestrian crossings.
(c) Waiting and loading restrictions.
(d) Obstruction and dangerous positions.
(e) Stopping to load and unload.
(f) Stopping at banned kerbs.
(g) Parking at night on roads.
(h) Overnight parking restrictions.
(i) Bus priority schemes.

3.2.2 Outline powers of bodies responsible for traffic control and enforcement.

Goods Syllabus

3.2.3 Explain the system for recovery of fixed penalty or excess parking charges.

3.3 Traffic offences and legal action (2 1/2 hours).

3.3.1 List offences involving automatic or discretionary disqualification, endorsement, fines or imprisonment.

3.3.2 Explain the procedure for disqualification.

3.3.3 State the procedure for, and implications of, endorsement of licences.

3.3.4 State the legal implications of being in charge of, or driving, a vehicle when under the influence of drink or drugs.

3.3.5 State the possible consequences of forgery and false statements.

3.3.6 State the liability of vehicle owners.

3.3.7 Describe the fixed penalty system.

A.4 Procedure in Case of Traffic Accidents (1 1/2 hours)

4.1 State the circumstances in which drivers have a legal liability to stop.

4.2 Explain the need for the driver to give particulars, and the occasions when he is required to produce his licence and/or insurance certificate.

Passenger Syllabus

3.2.3 Explain the system for recovery of fixed penalty or excess parking charges.

3.3 Traffic offences and legal action (2 1/2 hours).

3.3.1 List offences involving automatic or discretionary disqualification, endorsement, fines or imprisonment.

3.3.2 Explain the procedure for disqualification.

3.3.3 State the procedure for, and implications of, endorsement of licences.

3.3.4 State the legal implications of being in charge of, or driving, a vehicle when under the influence of drink or drugs.

3.3.5 Define and state consequences of forgery and false statements.

3.3.6 State the liability of vehicle owners.

3.3.7 Describe the fixed penalty system.

A.4 Procedure in Case of Traffic Accidents (1 1/2 hours)

4.1 State the circumstances in which drivers have a legal liability to stop.

4.2 Explain the need for the driver to give particulars, and the occasions when he is required to produce his licence and/or insurance certificate.

Goods Syllabus

4.3 State the liability of the owner of the vehicle to give identity of driver.

4.4 Explain the requirements for reporting accidents to the police.

A.5 Compulsory Vehicle and Passenger Insurance (2 hours)

5.1 State the legal obligations for Third Party risks and passengers.

5.2 Explain the method by which an operator may carry his own risk.

5.3 State the purpose and function of an insurance certificate.

5.4 Explain reasons for invalidation of policies.

5.5 State requirements about notification of accidents to insurers.

A.6 Safe Loading of Vehicle and Transit of Goods (2 1/2 hours).

6.1 State the main points in the Regulations and/or Codes of Practice concerned with safety of loads on vehicles.

Passenger Syllabus

4.3 State the liability of the owner of the vehicle to give identity of driver.

4.4 Explain the requirements for reporting accidents to the police.

4.5 State the main points to be included in a driver's accident report.

4.6 Outline the requirements to report accidents to the appropriate Traffic Commissioners.

A.5 Compulsory Vehicle and Passenger Insurance (2 hours)

5.1 State the legal obligations for Third Party risks and passengers.

5.2 Explain the method by which an operator may carry his own risk.

5.3 State the purpose and function of an insurance certificate.

5.4 Explain reasons for invalidation of policies.

5.5 State requirements about notification of accidents to insurers.

6.2 State the main provisions of current legislation on the authorization of special types of vehicle.

6.3 State the application of C & U Regulations as far as loading of vehicles is concerned.

6.4 State the basic legal requirements for the transport of:
(a) Hazardous and dangerous loads.
(b) Abnormal indivisible loads.
(c) Livestock and food.

B. Technical Standards and Aspects of Operation (14 hours).

B.1 Weight and Dimensions of Vehicles (4 hours).

1.1 Vehicle weights.

1.1.1 Define:
(a) Unladen weight (ULW).
(b) Gross vehicle weight (GVW).
(c) Permissible maximum weight (PMW).
(d) Gross train weight (GTW).
(e) Gross plated weight (GPW).
(f) Kerb-side weight.

1.2 Axle weights.

1.2.1 Differentiate between imposed weights and axle weight limits.

1.2.2 State and apply the formula for calculating axle weights and weight distributions.

B. Technical Standards and Aspects of Operation (11 1/2 hours).

B.1 Weight and Dimensions of Vehicles (1 1/2 hours).

1.1 Vehicle weights.

1.1.1 Define:
(a) Gross vehicle weight (GVW).
(b) Unladen weight (ULW).
(c) Permissible maximum weight (PMW).

1.2 Axle weights.

1.2.1 State axle weight limits.

Goods Syllabus

1.3 Maximum dimensions of the vehicle.

1.3.1 Define:
(a) Overall length and width.
(b) Overhang.

1.3.2 Apply regulations on projecting loads.

1.3.3 Apply regulations for overall dimensions of vehicle and load.

B.2 Vehicle Selection (1 hour)

2.1 Definition of different types of vehicle and their suitability.

2.1.1 Define motor car, heavy motor car, heavy goods vehicle, goods vehicle (small/medium/large).

2.1.2 Differentiate between rigid and articulated vehicle, trailer and semi-trailer, composite trailer.

2.1.3 State the key points to be taken into consideration when selecting a suitable vehicle for the safe transport of a load.

B.3 Mechanical Conditions (9 hours)

3.1 Plating and Testing

3.1.1 State the main provisions of the Plating and Testing Regulations with special reference to the following:
(a) Vehicles covered.
(b) Manufacturer's plate.
(c) First examination.

Passenger Syllabus

1.3 Maximum dimensions of the vehicle.

1.3.1 Define:
(a) Overall length and width.
(b) Overhang.

1.3.2 State the regulations for overall dimensions of vehicles.

B.2 Vehicle Selection (1 hour)

2.1 Definition of different types of vehicle and their suitability.

2.1.1 Define motor car, heavy motor car, public service vehicle, small/large passenger carrying vehicles, articulated bus, community bus, dual purpose vehicles (passenger/goods).

2.1.3 State the key points to be taken into consideration when selecting a suitable vehicle.

B.3 Mechanical Conditions (9 hours)

3.1 Certificate of Initial Fitness/Annual Inspection.

3.1.1 Describe and distinguish between the Certificate of Initial Fitness and Annual Inspection.

Goods Syllabus

(d) Annual examinations.
(e) Notifiable alterations.

3.2 Construction and Use Regulations

3.2.1 Outline the purpose of current legislation in connection with C & U and Type Approval Regulations.

3.2.2 State the main provisions of current legislation, with special reference to the following:
Construction and Use –
(a) Brakes.
(b) Braking efficiency.
(c) Speedometers.
(d) Mirrors.
(e) Safety glass.
(f) Windscreen wipers and washers.
(g) Audible warning devices.
(h) Fuel tanks.
(i) Silencers and noise.
(j) Smoke emission.
(k) Tyres.
(l) Towing.
(m) Side guards and under-run bumpers.

Passenger Syllabus

3.2 Construction and Use Regulations, PSV (Conditions of Fitness, Equipment and Use) Regulations.

3.2.1 Outline the purpose of current legislation in connection with Construction & Use, Type Approval and Conditions of Fitness Equipment and Use Regulations.

3.2.2 State the main provisions of current legislation, with special reference to the following:
Construction and Use –
(a) Brakes.
(b) Braking efficiency.
(c) Speedometers.
(d) Mirrors.
(e) Safety glass.
(f) Windscreen wipers and washers.
(g) Audible warning devices.
(h) Fuel tanks.
(i) Silencers and noise.
(j) Television.
(k) Smoke emission.
(l) Tyres.
(m) Towing.
(n) Turning circle.

Fitness –
(a) Stability.
(b) Suspension.
(c) Guard rails.
(d) Brakes.
(e) Steering gear.
(f) Fuel tanks, carburettors etc.
(g) Exhaust systems.

Goods Syllabus

3.3 Lighting and marking

3.3.1 Outline the major requirements for vehicle marking.
(a) Obligatory and non-obligatory lamps.
(b) Indicators and reflectors.
(c) Additional lamps, reflectors and markers required for certain vehicles and their loads.
(d) Use of lamps.

3.4 Operator's responsibility in relation to maintenance records and standards.

3.4.1 Describe the systems employed in planned maintenance:
(a) Defect reports.
(b) Keeping of records.
(c) Regular inspection and servicing.

3.4.2 Identify the operator's responsibility for the condition of a vehicle, the maintenance of which is contracted out.

Passenger Syllabus

(h) Locking of nuts.
(i) Body, steps, stairs etc.
(j) Windscreens.
(k) Ventilation.

Equipment –
(n) Fire extinguishers.
(o) First aid equipment.

Use –
(p) Obstruction of exits, entrances, gangways etc.
(q) Cleanliness and maintenance of body.

3.3 Lighting

3.3.1 Outline the major requirements for vehicle lighting.
(a) Obligatory and non-obligatory lamps.
(b) Indicators and reflectors.

3.4 Operator's responsibility in relation to maintenance records and standards.

3.4.1 Describe the systems employed in planned maintenance:
(a) Defect reports.
(b) Keeping of records.
(c) Regular inspection and servicing.

3.4.2 Identify the operator's responsibility for the condition of a vehicle, the maintenance of which is contracted out.

Goods Syllabus

3.5 Fleet inspection and roadside checks.

3.5.1 Outline the system of fleet inspection and roadside checks.

3.5.2 State the purpose of the relevant forms (GV9, GV160 and related forms).

3.5.3 State the procedure for prohibition notices and for their removal.

3.5.4 State the powers of police, certifying officers and authorised examiners.

C. **Access to the G.B. Market** (6 hours)

C.1 **Licensing (6 hours)**

1.1 The system of O-Licensing, in particular the role of the Licensing Authority.

1.1.1 Define a goods vehicle in the context of O-licensing.

1.1.2 State the procedure for application for, re-application (renewal) and variation of an O licence.

1.1.3 Define objections and appeals and the conditions in which they are applicable, and suitability of operating centres.

1.1.4 Explain the provisions of current legislation relating to:
(a) Compliance with hours, weight and speed regulations.
(b) Previous conduct of the applicant.
(c) Maintenance and finance.

Passenger Syllabus

3.5 Fleet inspection and roadside checks.

3.5.1 Outline the system of fleet inspection and roadside checks.

3.5.2 State the purpose of the relevant forms (PSV414 and other related forms).

3.5.3 State the procedure for prohibition notices and for their removal.

3.5.4 State the powers of police, certifying officers and authorised examiners.

C. **Regulation of Road Passenger Services** (7 hours)

C.1 **Legislation** (1 hour)

1.1 Existing legislation.

1.1.1 Explain the background to current legislation relating to statutory control of the industry including the relevant sections of the principal Road Traffic and Transport Acts.

(d) Operating centre.
(e) Plating and testing record.
(f) Lorry routes.

1.1.5 Define and give examples of refusal, revocation, suspension, curtailment and premature termination.

1.1.6 State the main provisions and purpose of the Goods Vehicle Operators (Qualifications) Regulations 1977 (as amended).

C.2 Licensing (6 hours)

The system of licensing and related procedures, in particular the role of the Traffic Commissioners.

2.1 PSV operator licensing.

2.1.1 State the procedures for application, variation and renewal of a PSV operator licence.

2.1.2 Explain the provisions of current legislation referring to:
(a) Certificate of initial fitness.
(b) Compliance with hours and speed regulations.
(c) Previous conduct of the applicant.
(d) Maintenance and finance.
(e) Criteria for refusal, revocation, suspension and curtailment.

2.1.3 State the main provisions as to qualifications for a PSV operator licence.

Goods Syllabus

D. Business and Financial Management (14 Hours)

D.1 Financial Management (5 hours)

1.1 Sources and uses of funds.

1.1.1 List the main sources and uses of short and long term funds.

1.2 The purpose of accounts and balance sheets and management information derived from accounting.

1.2.1 Explain the purpose and significance of the trading (operating) and profit and loss account and balance sheet.

1.2.2 Define debtors and creditors.

1.3 Capital.

1.3.1 Define and state the method of calculating:

Passenger Syllabus

2.2 Road service licensing.

2.2.1 Define the types of service for which a road service licence is required and any variations in applications, conditions and special dispensations.

2.2.2 State the circumstances in which express carriage services and contract carriage services may be operated.

2.2.3 Outline current legislation relating to other types of passenger transport operation (e.g. community transport services, school buses, car sharing and experimental areas).

D. Business and Financial Management (13 Hours)

D.1 Financial Management (4 hours)

1.1 Sources and uses of funds.

1.1.1 List the main sources and uses of short and long term funds.

1.2 The purpose of accounts and balance sheets and management information derived from accounting.

1.2.1 Explain the purpose and significance of the trading (operating) and profit and loss account and balance sheet.

1.2.2 Define debtors and creditors.

1.3 Capital.

1.3.1 Define and state the method of calculating:

Goods Syllabus

(a) Working capital (current) ratio.
(b) Liquidity (acid test) ratio.
(c) Working capital.
(d) Total capital employed.
(e) Return on capital employed.

1.4 Cash flow.

1.4.1 Define and differentiate between cash flow and cash budgeting.

1.4.2 Identify problems which may affect cash flow.

1.5 Purchasing and stock control.

1.5.1 State the financial implications.

1.5.2 State the factors to be taken into account.

Passenger Syllabus

(a) Working capital (current) ratio.
(b) Liquidity (acid test) ratio.
(c) Working capital.
(d) Total capital employed.
(e) Return on capital employed.

1.4 Cash flow.

1.4.1 Define and differentiate between cash flow and cash budgeting.

1.4.2 Identify problems which may affect cash flow.

D.2 Elements of Costing/Charging (5 hours)

2.1 Vehicle, fleet and route costing.

2.1.1 Describe the purpose and methods of costing.

2.1.2 State factors to be taken into account when costing.

2.1.3 State current recommendations on costing practices.

2.2 Basic fare and charging structures.

2.2.1 Describe factors to be taken into account and methods of calculating:

Goods Syllabus

D.2 Commercial Conduct of the Business (6 hours)

2.1 Vehicle costing.

2.1.1 Describe the purpose and methods of costing.

2.1.2 Differentiate between the various costs involved in operating a transport undertaking, in order to calculate rates and charges.

2.2 Methods of payment and collection of revenue.

2.2.1 Describe the methods of invoicing and payment.

Passenger Syllabus

(a) Fare structures.
(b) Charging for contract hire.

2.2.2 Describe procedures for dealing with fares revision.

2.3 Grants and Subsidies.

2.3.1 Describe the role of central and local government in respect of grants and subsidies.

2.3.2 Describe circumstances in which they are applicable.

D.3 Commercial Conduct of the Business (2 hours)

3.1 Service planning and schedules.

3.1.1 Describe the basic principles of constructing:
(a) Timetables.
(b) Vehicle schedules.
(c) Crew schedules.
(d) Crew rosters.

3.2 Revenue collection and protection.

3.2.1 Describe the various revenue collection systems.

3.2.2 Describe the importance and main methods of revenue protection.

3.2.3 Describe the purpose and use of the relevant documentation.

3.2.4 Describe paying in procedures.

Goods Syllabus

2.3 Documents and records involved in commercial practice.

2.3.1 State the purpose and use of the following:
(a) Estimate.
(b) Quotation.
(c) Order.
(d) Consignment note and delivery note.
(e) Invoice.
(f) Credit note.
(g) Debit note.
(h) Statement of account.

2.4 Banking and similar services.

2.4.1 List the main services provided by banks to industry.

D.3 Other Services (3 hours)

3.1 Advantages and scope of insurance.

3.1.1 Describe the economic and commercial advantages of the various types of insurance and the risks they cover.

3.1.2 Explain the main types of general insurance for:
(a) Fire.
(b) Theft.
(c) Fidelity guarantee.
(d) Goods in transit.
(e) Employer's liability.
(f) Public liability.
(g) CMR (Contrait Merchandise Routiers).

Passenger Syllabus

3.3 Documents and records involved in commercial practice.

3.3.1 State the purpose and use of the following:
(a) Estimate.
(b) Quotation.
(c) Order.
(d) Invoice.
(e) Credit note.
(f) Debit note.
(g) Statement of account.

3.4 Banking and similar services.

3.4.1 List the main services provided by banks to industry.

D.4 Other Services (2 hours)

4.1 Advantages and scope of insurance.

4.1.1 Describe the economic and commercial advantages of the various types of insurance and the risks they cover.

4.1.2 Explain the main types of general insurance for:
(a) Fire.
(b) Theft.
(c) Fidelity guarantee.
(d) Employer's liability.
(e) Public liability.

Goods Syllabus

3.2 Methods of operating.

3.2.1 Define the role of the freight forwarder.

3.2.2 Define a clearing house.

3.2.3 Define the role of the sub-contractor.

3.2.4 Define groupage.

E. Law (11 Hours)

E.1 Structure of Law (3 hours)

1.1 Law of business and carriage.

1.1.1 Explain the factors involved in making legally enforceable contracts with particular regard to conditions of carriage.

1.1.2 State the mutual duties and responsibilities of:
(a) Employer and agent.
(b) Employer and employee.
(c) Contractor and sub-contractor.

1.2 Operator's and employee's liability for civil and criminal acts.

1.2.1 Define in relation to road haulage operations:
(a) Negligence of operator or employee.
(b) Private nuisance.
(c) Occupier's liability under the Occupier's Liability Act 1957.

Passenger Syllabus

4.2 Methods of operating.

4.2.1 State advantages and disadvantages of co-ordinating and integrating different types of road passenger transport.

4.2.2 Define the terms joint services and sub-contracting.

E. Law (11 Hours)

E.1 Structure of Law (3 hours)

1.1 Law of business and carriage.

1.1.1 Explain the factors involved in making legally enforceable contracts with particular regard to conditions of carriage.

1.1.2 State the mutual duties and responsibilities of:
(a) Employer and agent.
(b) Employer and employee.
(c) Contractor and sub-contractor.

1.2 Operator's and employee's liability for civil and criminal acts.

1.2.1 Define in relation to road passenger operations:
(a) Negligence of operator or employee.
(b) Private nuisance.
(c) Occupier's liability under the Occupier's Liability Act 1957.

Goods Syllabus

1.2.2 Outline the limitations on the road carrier's power to restrict his legal liability.

1.2.3 Define in relation to road haulage operations:
(a) Public nuisance.
(b) Obstruction of the police in the execution of their duties.

E.2 Company Law (1 hour)

2.1 The formation and legal function of a business.

2.1.1 Outline procedures for formation and registration of a company.

2.1.2 Outline procedures for legal operation and liquidation of a company.

E.3 Social Legislation (6 hours)

3.1 Industrial Relations.

Passenger Syllabus

1.2.2 Outline the limitations on the road carrier's power to restrict his legal liability to passengers and their luggage.

1.2.3 Define in relation to road passenger operations:
(a) Public nuisance.
(b) Obstruction of the police in the execution of their duties.

E.2 PSV Conduct Regulations

2.1 Describe current legislation concerning:
(a) Conduct of drivers, conductors and passengers.
(b) Standing passengers.
(c) Lost property.
(d) School travel and concessionary travel.

2.2 State the current regulations concerning the powers of the police, crew members and officials in respect of passenger conduct.

E.3 Company Law (1 hour)

3.1 The formation and legal function of a business.

3.1.1 Outline procedures for formation and registration of a company.

3.1.2 Outline procedures for legal operation and liquidation of a company.

E.4 Social Legislation (6 hours)

4.1 Industrial Relations.

Goods Syllabus

3.1.1 State the main objectives of current legislation on:
(a) Trade union and labour relations.
(b) Terms and conditions of employment.
(c) Employment protection.
(d) Industrial training and employment.

3.1.2 Give examples of its principal application including codes of practice.

3.2 Social Security.

3.2.1 State the main objectives of current legislation on:
(a) Redundancy.
(b) Sickness, unemployment, industrial injury.
(c) Pensions.

3.2.2 Give examples of its principal application.

3.3 Discrimination.

3.3.1 State the main objectives of current legislation on:
(a) Equal pay.
(b) Sex discrimination.
(c) Disabled persons' employment.
(d) Race relations.

3.3.2 Give examples of its principal application.

Passenger Syllabus

4.1.1 State the main objectives of current legislation on:
(a) Trade union and labour relations.
(b) Terms and conditions of employment.
(c) Employment protection.
(d) Industrial training and employment.

4.1.2 Give examples of its principal application including codes of practice.

4.2 Social Security.

4.2.1 State the main objectives of current legislation on:
(a) Redundancy.
(b) Sickness, unemployment, industrial injury.
(c) Pensions.

4.2.2 Give examples of its principal application.

4.3 Discrimination.

4.3.1 State the main objectives of current legislation on:
(a) Equal pay.
(b) Sex discrimination.
(c) Disabled persons' employment.
(d) Race relations.

4.3.2 Give examples of its principal application.

Goods Syllabus

3.4 Safety.

3.4.1 State the main objectives of the current legislation on health, safety, and welfare, in relation to employers, employees and the general public.

3.4.2 Give examples of its principal application, including codes of safe working practice.

E.4 Taxation (1 hour)

4.1 Vehicle excise duty and registration.

4.1.1 List the main points of the current Vehicles Excise and Finance Acts relevant to road freight transport operation.

4.2 Trade licences (plates).

4.2.1 State the circumstances in which vehicles may be used under trade licences instead of the more normal excise licence.

4.2.2 Describe the requirements involved in registration and use of trade licences (plates).

Passenger Syllabus

4.4 Safety.

4.4.1 State the main objectives of the current legislation on health, safety, and welfare, in relation to employers, employees and the general public.

4.4.2 Give examples of its principal application, including codes of safe working practice.

E.5 Taxation (1 hour)

5.1 Vehicle excise duty and registration.

5.1.1 List the main points of the current Vehicles Excise and Finance Acts relevant to road passenger transport operation.

5.2 Trade licences (plates).

5.2.1 State the circumstances in which vehicles may be used under trade licences instead of the more normal excise licence.

5.2.2 Describe the requirements involved in registration and use of trade licences (plates).

Appendix 4 ~ Specimen Questions

<u>Special Note</u> These questions fall into three groups:- Questions 1-8 are for National Goods candidates, Questions 9-16 are for both National Goods and National PSV candidates, whilst Questions 17-24 are for National PSV candidates. Those readers preparing for the CPC in National Goods operation should therefore be able to answer Questions 1-16, whilst National PSV candidates should be able to answer Questions 9-24.

State which of the alternative answers to each question is correct.

National Goods

1. A road freight operator conducts his business from his home, garages his vehicles at the premises of his major customer, and has a contract with a local garage for their maintenance. His Operating Centre is:-
(a) his home; (b) the garage; (c) the customer's premises; (d) none of the above.

2. A Standard Goods Vehicle Operator's Licence is required by a manufacturer making deliveries of his product with vehicles over 3.5 tonnes Gross Plated Weight.
(a) True; (b) False.

3. If a Professional Competent full-time Transport Manager designated by a Goods Operator resigns, the operator's Standard O-Licence will continue in force for up to:-
(a) 6 months; (b) 9 months; (c) 12 months; (d) 15 months.

4. A vehicle may be driven whilst laden on 'Trade Plates' in order to demonstrate it to a prospective customer.
(a) True; (b) False.

5. A vehicle examiner can require the driver of a Goods Vehicle to proceed to a location where the vehicle can be tested or weighed. This location is not more than:-
(a) 5 miles distant; (b) 1 mile distant; (c) 5 kilometres distant; (d) 10 kilometres distant.

6. A vehicle may be plated below the Gross Plated Weight shown in the Standard List if:-
(a) The vehicle cannot meet the required braking standard at this weight; (b) The operator fits tyres of a lower ply rating than standard; (c) There is a change of use of the vehicle; (d) The operator agrees to limit its maximum gross weight to the new Gross Plated Weight.

7. Defect reports from drivers must be retained for a minimum period of:-
(a) 6 months; (b) 9 months; (c) 12 months; (d) 15 months.

8. The maximum permitted speed of a rigid goods vehicle over 7.5 tonnes Gross Plated Weight on Motorways is:-
(a) 40 m.p.h.; (b) 50 m.p.h.; (c) 60 m.p.h.; (d) 70 m.p.h.

National Goods and National PSV

9. If a driver is involved in an accident which causes damage to another vehicle, and for any reason does not give certain particulars to any person reasonably requiring these, he must report the accident to the Police as soon as possible, and in any case within:-
(a) 12 hours; (b) 24 hours; (c) 48 hours; (d) 5 days.

10. A double yellow line painted on, and at right angles to, the kerb means:-
(a) No parking during the working day; (b) No loading or unloading during the working day; (c) No parking at any time; (d) No loading or unloading at any time.

11. Completed tachograph discs must be kept for not less than:-
(a) 6 months; (b) 9 months; (c) 12 months; (d) 15 months.

12. The Braking Efficiency standard required of the main brake of a newly-registered vehicle is:-
(a) 16%; (b) 25%; (c) 45%; (d) 50%.

13. The maximum period of continuous driving permitted before a break must be taken by a driver on National journeys and work is:-
(a) 4 hours; (b) 5 hours; (c) 5 hours 30 minutes; (d) 8 hours 30 minutes.

14. Which of the following is a running cost?
(a) Licences; (b) Excise Duty (Road Tax); (c) Fuel; (d) Insurance.

15. If a carrier retains the goods of his customer because the latter owes him money for his freight charges, he is said to be exercising his right of:-
(a) Bailment; (b) Lien; (c) Demurrage; (d) Stoppage in transit.

16. The maximum permitted Gross Plated Weight of a rigid 4-wheel vehicle is:-
(a) 10 tons; (b) 14 tons; (c) 16 tons; (d) 18 tons.

National PSV

17. The maximum permitted speed on a de-restricted dual carriageway for a public service vehicle with 52 seats is:-
(a) 70 m.p.h.; (b) 60 m.p.h.; (c) 50 m.p.h.; (d) 40 m.p.h.

18. A motor vehicle used to carry passengers for hire and reward, but not at separate fares, will be classed as a Public Service Vehicle if it has seats for:-
(a) 4 or more passengers; (b) 8 or more passengers; (c) 9 or more passengers; (d) 16 or more passengers.

19. The minimum age for driving a double-deck public service vehicle is:-
(a) 17 years; (b) 18 years; (c) 21 years; (d) 25 years.

20. A road service licence is required to operate:-
(a) a contract carriage; (b) an express carriage; (c) a stage carriage; (d) an excursion and tour.

21. The maximum spreadover allowed for a public service vehicle driver on Domestic work is:-
(a) 12 hours; (b) 14 hours 30 minutes; (c) 13 hours; (d) 16 hours.

22. A public service vehicle driver on a Regular Service may work a maximum continuous duty (without a 30 minute break) of:-
(a) 5 hours 30 minutes; (b) 8 hours 30 minutes; (c) 8 hours; (d) 4 hours.

23. The maximum permitted height of a public service vehicle is:-
(a) 4.57 metres; (b) 4.00 metres; (c) 4.20 metres; (d) 3.87 metres.

24. In respect of personal injuries to fare-paying passengers, the Public Passenger Vehicles Act 1981 imposes on PSV operators:-
(a) strict liability; (b) liability for negligence only; (c) absolute liability; (d) criminal liability.

Answers will be found on page 257 (inside back cover).

Index